10 Indiana ILEARN Grade 8 Math Practice Tests

The Ultimate Test Prep Collection with Answer Explanations

Dr. A. Nazari

10 Practice Tests

 *Grade 8 Mathematics* 

Welcome!

This book contains **10 full-length practice tests** — the most comprehensive preparation you can get for your Grade 8 math assessment. Each test covers all six topics:

📖 *Irrational Numbers*   📖 *Powers & Scientific Notation*

📖 *Linear Equations*   📖 *Functions*

📖 *Geometry*   📖 *Data & Relationships*

Ten tests give you the practice needed to walk into the real test feeling fully prepared.

Thorough preparation leads to outstanding results.

> **“** *Ten full tests! By the time you finish, there won't be any surprises on test day.* **”**

 How to Use This Book

What's Inside

- **10 Full-Length Practice Tests** — each covers all 6 chapters of Grade 8 math: irrational numbers, exponents & scientific notation, linear equations, functions, geometry, and data analysis.
- **Detailed Answer Explanations** — every question includes a step-by-step solution so you learn from every mistake.
- **Formula Reference Sheet** — all the key Grade 8 formulas you need, organized and ready for quick review.
- **Test Tracker** — log your scores across all 10 tests and monitor your progress from start to finish.

Your 10-Test Training Plan

★ PHASE 1: Foundation (Tests 1–3)

Untimed or soft-timed. Focus on understanding the format, identifying strengths and weaknesses, and building good study habits.

★★ PHASE 2: Building Skills (Tests 4–7)

Timed (70 minutes each). Work on pacing, accuracy, and showing complete solutions. Review weak topics between tests.

★★★ PHASE 3: Test-Day Ready (Tests 8–10)

Full test conditions: strict timing, quiet space, no notes. Compare scores with your early tests to see your growth.

Schedule: Take one test every 3–4 days, or one per week. Use study days between tests to review.

Types of Questions

● **Multiple Choice:** *Four options — work the problem first, then match. Eliminate obviously wrong answers to narrow your choices.*

Short Answer & Constructed Response: *Show every step: equations, substitutions, simplifications. Partial credit rewards correct reasoning even if the final answer is off.*

Graphing & Data Analysis: *Plot points, draw lines, interpret graphs. Label axes clearly.*

Tip: Ten tests is a full preparation program. Don't rush. The key is what you do between tests — study, review, and understand your mistakes before moving forward.

Find more at
ViewMath.com/IN-Grade8

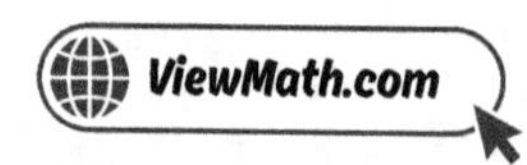

Test-Taking Tips
Your complete test-day toolkit

Before the Test

- Review your notes from the previous test — focus on your weak topics
- Set up a quiet, clean workspace with all your materials ready
- Start with a positive mindset: you've prepared for this

During the Test

- Read each problem fully before calculating anything
- Write the formula or set up the equation first, then substitute values
- Show all your work — every step, every operation
- If stuck for more than 2 minutes, mark it and move on
- Use estimation to check if your answers are reasonable

After the Test

- Read the full explanation for every question you got wrong
- Write down which topics gave you trouble (not just question numbers)
- Study those topics before taking the next test
- Record your score in the Test Tracker

⚠ Common Mistakes in Grade 8 Math

⚠ **Exponents:** $(ab)^n = a^n b^n$, but $a^m + a^n \neq a^{m+n}$. Only multiply/divide to combine.

⚠ **Slope formula:** $m = \frac{y_2 - y_1}{x_2 - x_1}$ — keep the order consistent.

⚠ **Systems of equations:** The solution must satisfy both equations.

⚠ **Transformations:** Rotations and reflections change position; dilations change size.

⚠ **Volume:** Use $\pi \approx 3.14$ or leave as π — match what the question asks.

The students who improve the most aren't the ones who take the most tests — they're the ones who carefully review every mistake. Make that your priority.

Find more at
ViewMath.com/IN-Grade8

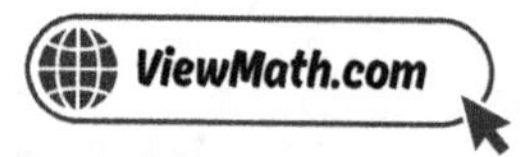

 What You'll Need

Materials Checklist

Sharpened Pencils — #2 pencils, at least two

Good Eraser — for clean corrections

Scratch Paper — for working out problems

Ruler / Straightedge — for graphing & geometry

Quiet Space — no distractions

Focused Mind — ready to do your best

Allowed Materials

- ✔ Pencils and eraser
- ✔ Scratch paper (provided on official test day)
- ✔ Ruler or straightedge (if required)
- ✔ Protractor (if required)

Not Allowed

- ✘ Calculator (unless your state test allows it)
- ✘ Cell phone or any electronic device
- ✘ Notes, textbooks, or reference sheets
- ✘ Help from others during the test

♥ A Note for Parents & Guardians

Ten tests is a comprehensive program. Plan **one test every 3–4 days** *(or one per week) with study sessions between each test.*

How to help:

- *Tests 1–3 should be untimed — build understanding before adding pressure.*
- *After each test, review the answer explanations together. Ask: "Which topics were hardest? Let's study those before the next one."*
- *Use the Test Tracker to celebrate progress over time.*
- *For topic-specific help, pair this book with our* **Grade 8 Math Study Guide** *or* **Grade 8 Workbook**.

Find more at
ViewMath.com/IN-Grade8

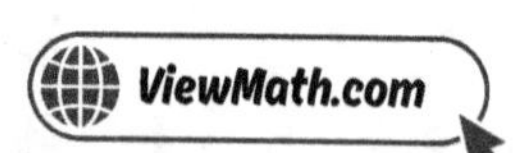

Grade 8 Formula Reference

Keep this page handy — you may use it during your practice tests!

X^1 Exponent Rules

$$a^m \cdot a^n = a^{m+n} \qquad (a^m)^n = a^{mn} \qquad (ab)^n = a^n \cdot b^n$$

$$\frac{a^m}{a^n} = a^{m-n} \qquad a^0 = 1 \ (a \neq 0) \qquad a^{-n} = \frac{1}{a^n}$$

Lines & Linear Equations

Slope: $m = \dfrac{y_2 - y_1}{x_2 - x_1} = \dfrac{rise}{run}$

m = slope b = y-intercept

Slope-intercept: $y = mx + b$

Parallel lines: same slope

Proportional: $y = mx$

Proportional: passes through origin

Scientific Notation

$a \times 10^n$ where $1 \leq |a| < 10$

Multiply: add exponents

Divide: subtract exponents

$\sqrt{x}$ Roots & Number Sense

Perfect squares: 1, 4, 9, 16, 25, 36, 49, 64, 81, 100, 121, 144

Perfect cubes: 1, 8, 27, 64, 125

$$\sqrt{2} \approx 1.414 \qquad \sqrt{3} \approx 1.732 \qquad \pi \approx 3.14159$$

Pythagorean Theorem & Distance

$$a^2 + b^2 = c^2 \qquad c = \text{hypotenuse (longest side of a right triangle)}$$

Distance: $d = \sqrt{(x_2 - x_1)^2 + (y_2 - y_1)^2}$

Volume Formulas

Cylinder $V = \pi r^2 h$ **Cone** $V = \dfrac{1}{3}\pi r^2 h$ **Sphere** $V = \dfrac{4}{3}\pi r^3$

⬚ Angle Relationships

Triangle angle sum: 180°   **Exterior angle** = sum of two remote interior angles

Parallel lines + transversal: Alternate interior angles are equal · Co-interior angles sum to 180°

⬚ Functions

Each input → exactly one output   **Vertical line test:** if any vertical line hits graph more than once ⇒ not a function

Linear: constant rate of change ($y = mx + b$)   **Nonlinear:** rate of change varies

↻ Transformations

Translation: slide   **Reflection:** flip   **Rotation:** turn   **Dilation:** resize

Congruent = same shape & size   Similar = same shape, proportional size

Tip: Bookmark this page! Review it before each test so these formulas become second nature.

Get Online

Find more at
ViewMath.com/IN-Grade8

⬛ Multiplication Table ⬛

You may use this table during your practice tests!

×	1	2	3	4	5	6	7	8	9	10	11	12
1	1	2	3	4	5	6	7	8	9	10	11	12
2	2	4	6	8	10	12	14	16	18	20	22	24
3	3	6	9	12	15	18	21	24	27	30	33	36
4	4	8	12	16	20	24	28	32	36	40	44	48
5	5	10	15	20	25	30	35	40	45	50	55	60
6	6	12	18	24	30	36	42	48	54	60	66	72
7	7	14	21	28	35	42	49	56	63	70	77	84
8	8	16	24	32	40	48	56	64	72	80	88	96
9	9	18	27	36	45	54	63	72	81	90	99	108
10	10	20	30	40	50	60	70	80	90	100	110	120
11	11	22	33	44	55	66	77	88	99	110	121	132
12	12	24	36	48	60	72	84	96	108	120	132	144

💡 How to Use This Table

To find **4 × 7**:

1. Find **4** in the left column (blue).
2. Find **7** in the top row (blue).
3. Follow the row and column until they meet: the answer is **28**!

> **ℹ Tip:** You can also use this table for division! If you know $28 \div 4 =$?, find 28 in the 4's row. The column header gives you the answer: **7!**

Find more at
ViewMath.com/IN-Grade8

📈 My Test Tracker 📈

Record every test and watch your scores improve

Name: _______________________________   **Start Date:** ___________________

GETTING STARTED (Tests 1–3)

Test 1 — Untimed

Date: ____________   Score: ______ / ______   %: ______   Topics to review: ____________________

Test 2 — Untimed

Date: ____________   Score: ______ / ______   %: ______   Topics to review: ____________________

Test 3 — Soft Timer

Date: ____________   Score: ______ / ______   %: ______   Topics to review: ____________________

BUILDING SKILLS (Tests 4–7)

Test 4 — Timed (70 min)

Date: ____________   Score: ______ / ______   %: ______   Focus area: ____________________

Test 5 — Timed (70 min)

Date: ____________   Score: ______ / ______   %: ______   Focus area: ____________________

Test 6 — Timed (70 min)

Date: ____________   Score: ______ / ______   %: ______   Focus area: ____________________

Test 7 — Timed (70 min)

Date: ____________   Score: ______ / ______   %: ______   Focus area: ____________________

TEST-DAY READY (Tests 8–10)

Test 8 — Full Test Conditions

Date: _____________   Score: ________ / ________   %: ________   Growth since Test 1: _______________

Test 9 — Full Test Conditions

Date: _____________   Score: ________ / ________   %: ________   Growth since Test 1: _______________

Test 10 — Full Test Conditions

Date: _____________   Score: ________ / ________   %: ________   Growth since Test 1: _______________

📊 Score Progress

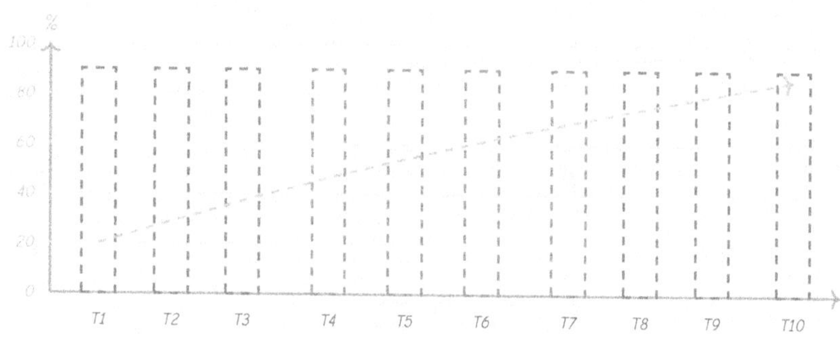

Shade each bar after every test. Watch your improvement!

✅ Final Reflection

The most important thing I learned: ____

The topic where I improved the most: ____

My advice for other students: ____

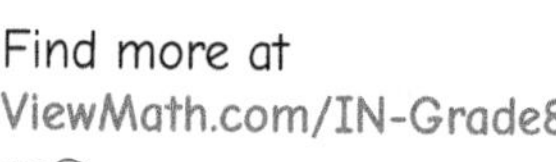

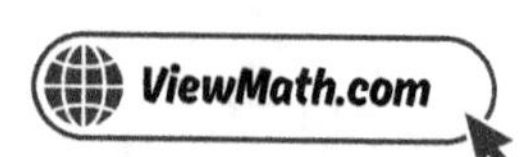

⭐ Table of Contents ⭐

Here's what we'll explore together!

Let's learn and have fun!

1

Practice Test 1

📋 30 Questions

✏️ Before You Start ✏️

- ✔ **Read each question carefully** before choosing your answer.
- ✔ **Show your work** on scratch paper when you need to.
- ✔ **Skip hard questions** and come back to them later.
- ✔ **Check your answers** when you're done.
- ✔ **Take your time** — there's no rush!

⭐ You've Got This! ⭐

Do your best and show what you know!

1. *True or false: The number $\frac{22}{7}$ is equal to π.*

   Your Answer:

2. *The number line below shows $\sqrt{n}$ placed between two tick marks. Based on the position, what could n be?*

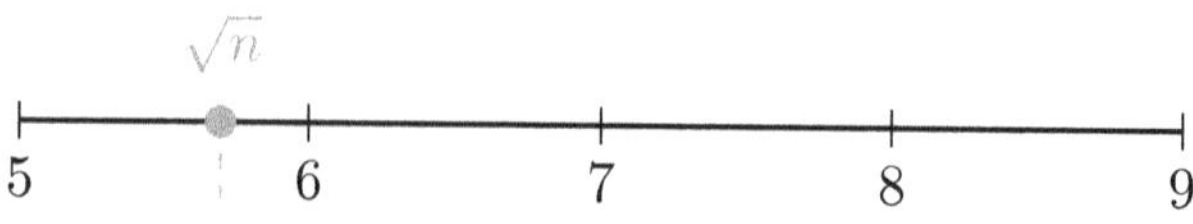

   (A) $n = 28$

   (B) $n = 33$

   (C) $n = 40$

   (D) $n = 55$

3. *The diagram below shows a rectangle with irrational side lengths. Which is the best estimate of the area?*

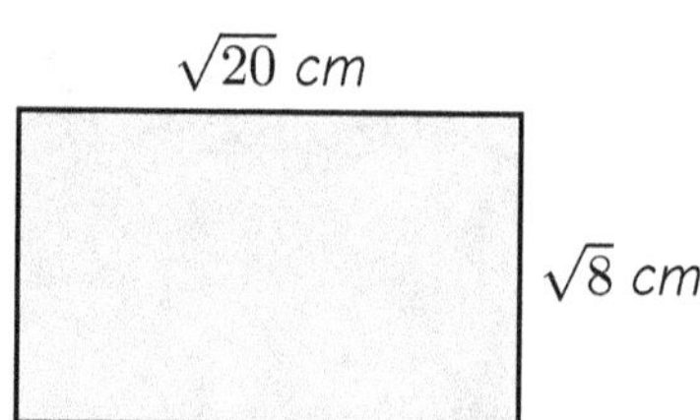

   (A) $\sqrt{28} \approx 5.3\ cm^2$

   (B) $\sqrt{160} \approx 12.6\ cm^2$

   (C) $28\ cm^2$

   (D) $160\ cm^2$

4. *Simplify $7^5 \cdot 7^{-5}$.*

   (A) 7^{25}

   (B) 7^{10}

   (C) 0

   (D) 1

5. The diagram shows a square and a cube. The square has an area of A square units and the cube has a volume of V cubic units. If $A = V$, which pair of side lengths is correct?

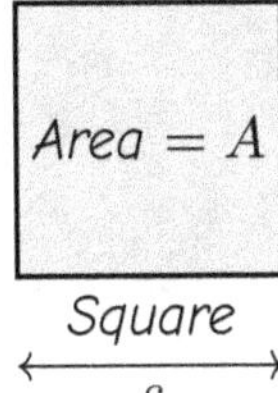

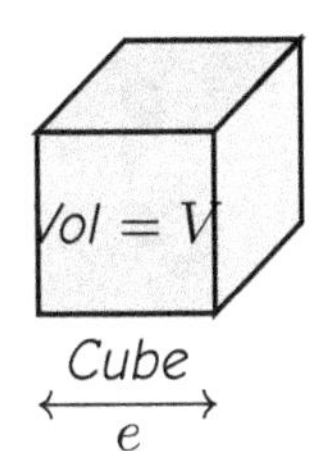

(A) $s = 6,\ e = 6$ (B) $s = 8,\ e = 4$

(C) $s = 9,\ e = 3$ (D) $s = 5,\ e = 5$

6. Earth's mass is about 6×10^{24} kg and Jupiter's mass is about 1.9×10^{27} kg. About how many times more massive is Jupiter than Earth? Round to the nearest whole number.

Your Answer:

7. The table and the equation below each describe a proportional relationship.

x	1	2	3	4
y	6	12	18	24

Equation: $y = 5x$

Which relationship has the larger constant of proportionality?

(A) The table (B) The equation

(C) They are equal. (D) Not enough information.

8. A phone plan charges $0.15 per text. If you graph total cost vs. number of texts, the slope is:

(A) 15 (B) 0.15

(C) 1.50 (D) 0.015

Find more at
ViewMath.com/IN-Grade8

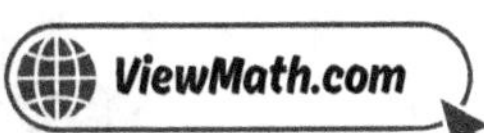

9. A gym charges a $25 joining fee plus $10 per month. Write an equation for the total cost y after x months.

Your Answer:

10. A school play sold 200 tickets. Adult tickets were $8 and student tickets were $5. Total revenue was $1,240. How many adult tickets were sold?

Your Answer:

11. Use the graph of g below. For which input is $g(x) = 0$?

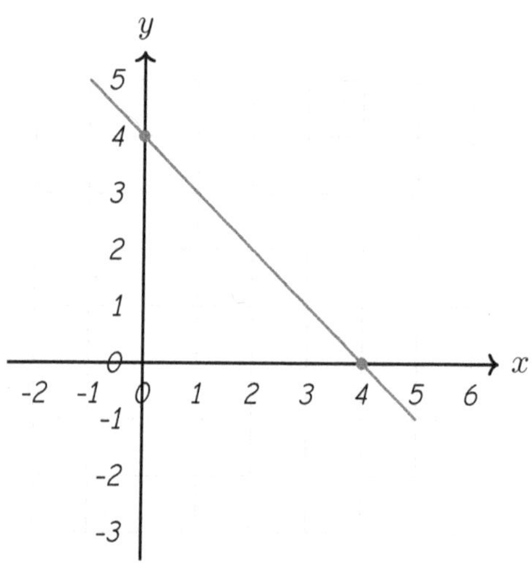

(A) $x = 0$

(B) $x = 2$

(C) $x = 4$

(D) $x = -4$

12. The tables below show two functions. Find the rate of change and initial value of each. Which function will have a greater value at $x = 10$?

<table>
<tr><td colspan="2" align="center">Function A</td></tr>
<tr><td>x</td><td>y</td></tr>
<tr><td>0</td><td>20</td></tr>
<tr><td>1</td><td>23</td></tr>
<tr><td>2</td><td>26</td></tr>
<tr><td>3</td><td>29</td></tr>
</table>

<table>
<tr><td colspan="2" align="center">Function B</td></tr>
<tr><td>x</td><td>y</td></tr>
<tr><td>0</td><td>5</td></tr>
<tr><td>1</td><td>10</td></tr>
<tr><td>2</td><td>15</td></tr>
<tr><td>3</td><td>20</td></tr>
</table>

Your Answer:

13. Does the equation $y = -12$ represent a linear function? Write Yes or No.

Your Answer:

14. A taxi charges a \$4 flat fee plus \$3 per mile. Which equation models the total cost y for x miles?

(A) $y = 4x + 3$

(B) $y = 3x + 4$

(C) $y = 7x$

(D) $y = 3x - 4$

15. A graph shows a line going upward with a constant slope. Is the rate of change constant or varying?

Your Answer:

16. Which of the following is NOT preserved under a translation?

(A) Side lengths

(B) Angle measures

(C) Position

(D) Parallelism of sides

Find more at
ViewMath.com/IN-Grade8

17. *Describe a sequence of rigid transformations that maps a triangle with vertices $(1,1)$, $(4,1)$, $(1,3)$ to a triangle with vertices $(-1,1)$, $(-4,1)$, $(-1,3)$.*

Your Answer:

18. *Point $(5,2)$ is rotated $180°$ around the origin. What are the coordinates of the image?*

Your Answer:

19. *A dilation centered at the origin maps $(4,-2)$ to $(12,-6)$. What is the scale factor?*

(A) 2

(B) 3

(C) 4

(D) 6

20. *Parallel lines ℓ and m are cut by transversal t. If a pair of alternate interior angles are $(4x)°$ and $80°$, what is x?*

(A) 320

(B) 20

(C) 25

(D) 10

21. *A right triangle has legs 5 and 5. What is the hypotenuse?*

(A) 10

(B) $\sqrt{50}$

(C) 25

(D) $\sqrt{10}$

22. *What is the distance between $(2,-5)$ and $(6,-2)$?*

(A) 25

(B) 7

(C) 5

(D) $\sqrt{7}$

23. A cylinder has a volume of 450π cm^3. A cone has the same base and height. What is the cone's volume?

(A) 150π cm^3

(B) 225π cm^3

(C) 1350π cm^3

(D) 900π cm^3

---

24. How many square pyramids with base side 4 cm and height 6 cm could fill a rectangular prism with dimensions $4 \times 4 \times 6$ cm?

(A) 1

(B) 2

(C) 3

(D) 4

---

25. As temperature increases, heating costs decrease. This relationship is:

(A) a positive linear association

(B) a negative linear association

(C) no association

(D) a nonlinear association

---

26. A scatter plot shows these points with a trend line through $(1, 3)$ and $(5, 7)$. What is the equation of the trend line?

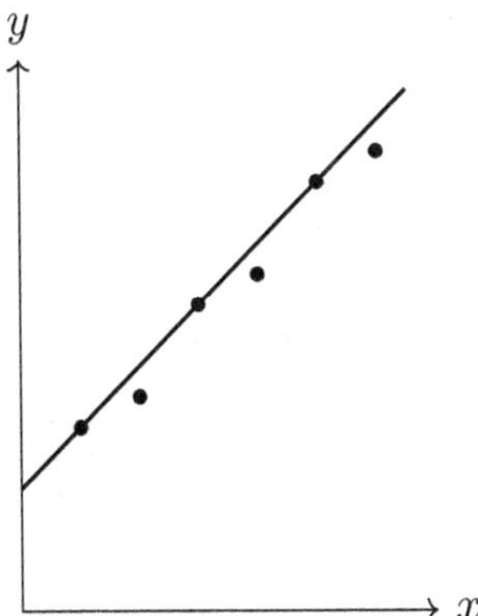

(A) $y = x + 2$

(B) $y = 2x + 1$

(C) $y = x + 3$

(D) $y = 0.5x + 2.5$

27. *Why can extrapolation be unreliable?*

(A) Because you cannot use a calculator

(B) Because the pattern may not continue outside the observed data range

(C) Because the slope is always wrong

(D) Because the y-intercept is always zero

28. *Using the table from q19, what percentage of boys prefer reading? What percentage of girls prefer reading? Is there an association?*

Your Answer:

29. $P(event) = \frac{3}{4}$. *What is* $P(not\ event)$?

(A) $\frac{3}{4}$

(B) $\frac{1}{4}$

(C) $\frac{1}{2}$

(D) 0

30. *Explain why* $\binom{5}{2} = \binom{5}{3}$.

Your Answer:

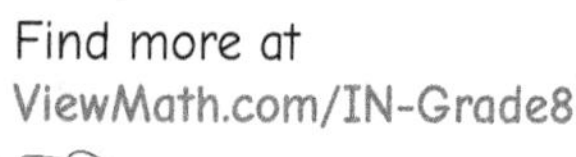

 # End of Practice Test 1 

Great job finishing the test!

 My Score

I got _____________ out of 30 questions right.

Check your answers in the **Answer Key** at the back of the book.

Review any questions you missed. That's how we learn!

Check Your Score Online!

Visit **ViewMath Academy** to enter your answers and see which topics you need to review. You can also explore lessons, take quizzes, track your scores, and save your progress!

viewmath.com/score/8.1.IN.16

Or go to viewmath.com/score and enter code: 8.1.IN.16

Practice Test 2

 30 Questions

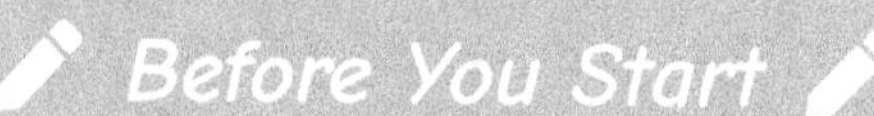

✏ Before You Start ✏

- ✓ **Read each question carefully** before choosing your answer.
- ✓ **Show your work** on scratch paper when you need to.
- ✓ **Skip hard questions** and come back to them later.
- ✓ **Check your answers** when you're done.
- ✓ **Take your time** — there's no rush!

⭐ **You've Got This!** ⭐

Do your best and show what you know!

1. The number line below shows two points, P and Q.

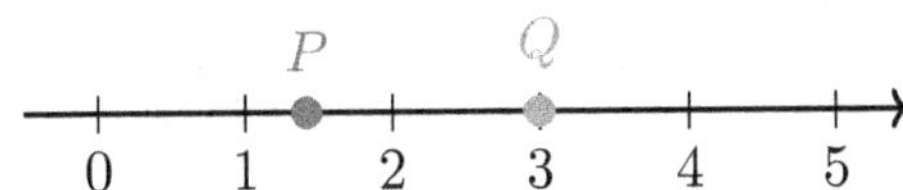

Point P represents $\sqrt{2}$ and point Q represents 3. Which statement is true?

(A) Both P and Q are rational.

(B) Both P and Q are irrational.

(C) P is irrational and Q is rational.

(D) P is rational and Q is irrational.

2. On the number line below, each tick mark represents 0.1 units. Plot the approximate location of $\sqrt{5}$ and label it.

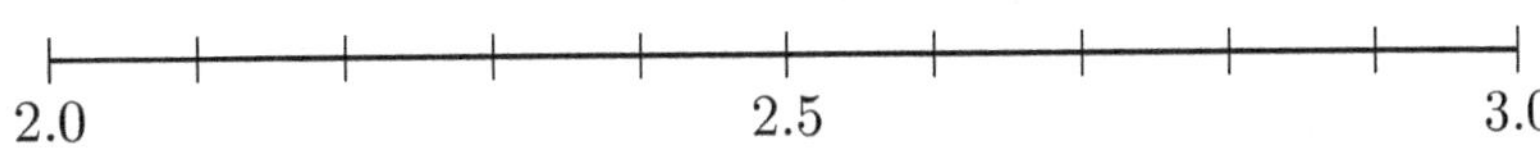

Your Answer:

3. The number line below shows the values of π and $\sqrt{10}$ marked. Find the distance between them.

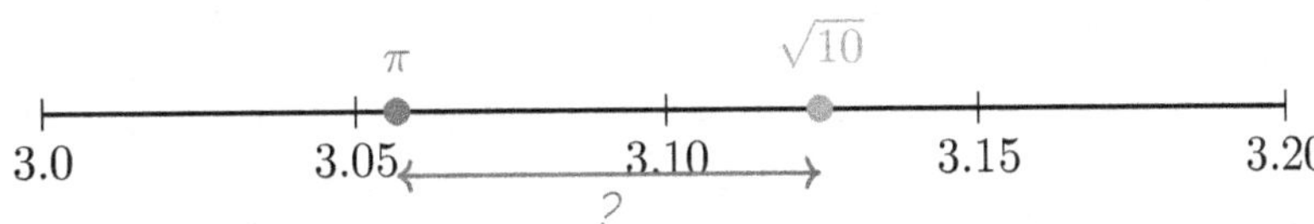

Your Answer:

4. Which expression is equivalent to 4^{-3}?

(A) -64

(B) -12

(C) $\frac{1}{12}$

(D) $\frac{1}{64}$

5. Solve $x^2 = \frac{9}{16}$.

  (A) $x = \frac{3}{4}$ only   (B) $x = \pm\frac{3}{4}$

  (C) $x = \frac{9}{8}$ (D) $x = \pm\frac{9}{4}$

6. What is $9.1 \times 10^5 - 8.6 \times 10^5$?

  (A) 0.5×10^5 (B) 5×10^4

  (C) 5×10^5 (D) Both A and B

7. A graph of a proportional relationship passes through the point $(5, 20)$. What is the constant of proportionality k?

  (A) 5   (B) 15

  (C) 4   (D) 100

8. A jogger runs 2 miles in 20 minutes and 5 miles in 50 minutes. What is the slope of distance vs. time (in miles per minute)?

Your Answer

9. A line passes through $(1, 2)$ and $(4, 11)$. What is the equation in slope-intercept form?

  (A) $y = 3x - 1$ (B) $y = 3x + 1$

  (C) $y = 3x + 2$ (D) $y = 3x - 2$

10. You have \$5 bills and \$10 bills totaling \$85. You have 12 bills. How many \$10 bills do you have?

Your Answer

Find more at
ViewMath.com/IN-Grade8

11. The function $f(x) = 10 - 3x$ gives the number of pieces of candy remaining after x friends each take 3. How many pieces are left after 2 friends take candy?

(A) 1

(B) 4

(C) 7

(D) 16

12. A table for Function C shows $(0, 7)$, $(1, 11)$, $(2, 15)$. What is the initial value?

Your Answer:

13. Look at the graph below. Is the function linear or nonlinear? Explain your reasoning.

Your Answer:

14. The table shows $(2, 9)$ and $(6, 25)$. Find the slope of the line.

Your Answer:

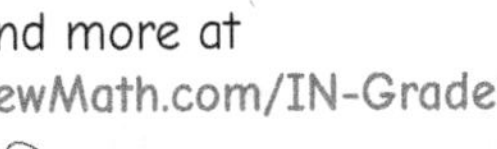

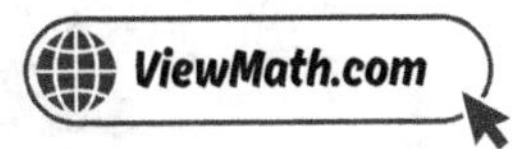

15. A graph is a straight line segment going upward. The function during this segment is:

(A) Nonlinear and increasing

(B) Linear and decreasing

(C) Linear and increasing

(D) Nonlinear and constant

16. A figure is reflected over the y-axis. What happens to point $(7, -2)$?

(A) $(7, 2)$

(B) $(-7, 2)$

(C) $(-7, -2)$

(D) $(-2, 7)$

17. Triangle P has angles $40°$, $60°$, and $80°$. Triangle Q has angles $40°$, $60°$, and $80°$. Are they congruent?

(A) Yes, because matching angles guarantee congruence.

(B) No, because angles alone do not guarantee congruence.

(C) Yes, because all three angles match.

(D) No, because the angles are in different order.

18. A point is translated by $(3, -2)$ and then reflected over the y-axis. If the original point is $(1, 4)$, what is the final image?

(A) $(-4, 2)$

(B) $(4, 2)$

(C) $(-4, -2)$

(D) $(4, -2)$

19. Two similar rectangles have a scale factor of 3. The smaller has area 12 cm^2. What is the area of the larger?

Your Answer:

20. Two parallel lines are cut by a transversal. One angle measures $72°$. What is its alternate interior angle?

(A) $18°$

(B) $72°$

(C) $108°$

(D) $288°$

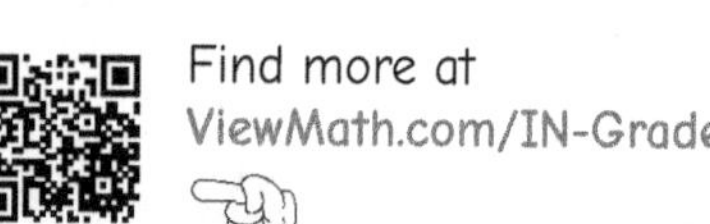

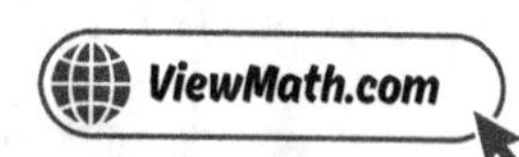

21. A right triangle has legs 8 and 15. What is the hypotenuse?

(A) 23

(B) 17

(C) $\sqrt{23}$

(D) 289

22. Find the distance between $(5, -2)$ and $(5, 7)$.

Your Answer:

23. What is the volume of a cylinder with radius 7 m and height 3 m? Leave your answer in terms of π.

(A) $21\pi \ m^3$

(B) $63\pi \ m^3$

(C) $147\pi \ m^3$

(D) $49\pi \ m^3$

24. A triangular pyramid has a base with legs 5 cm and 12 cm (a right triangle) and a height of 10 cm. Find the volume.

Your Answer:

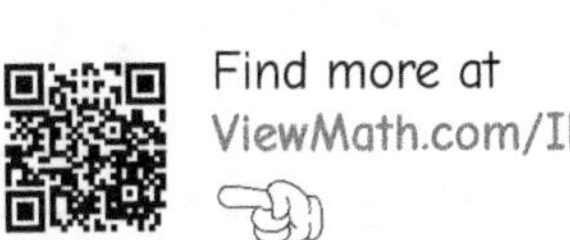

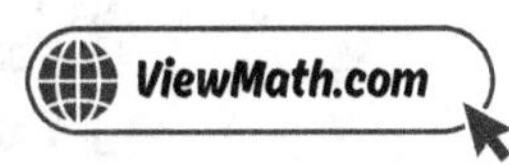

25. *Identify the outlier in the scatter plot.*

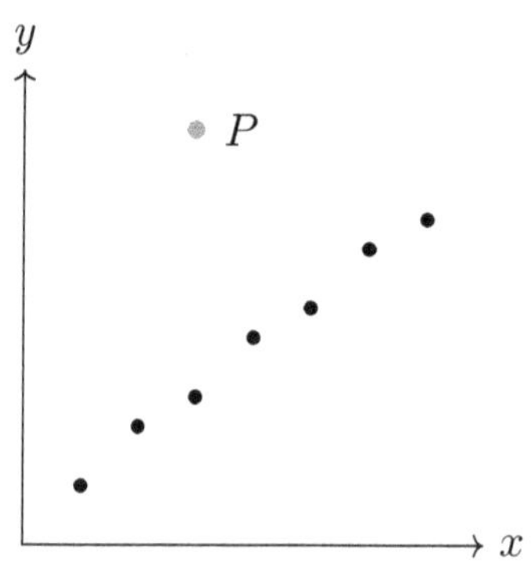

(A) $(1, 1)$

(B) $(7, 5.5)$

(C) $(3, 7)$ — point P

(D) $(4, 3.5)$

26. *Draw a line of best fit through the data and estimate its slope.*

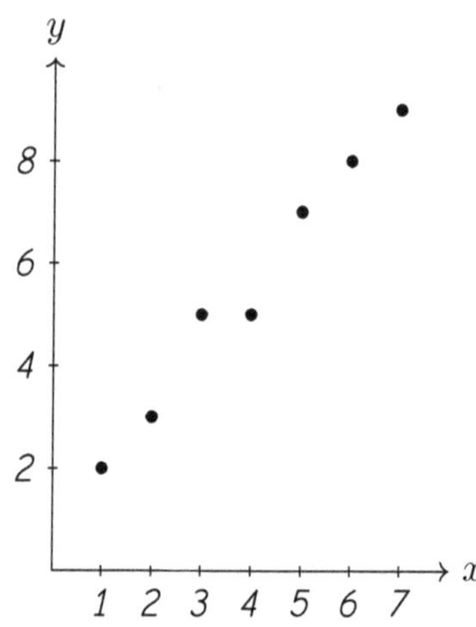

Your Answer:

27. *A linear model is* $y = 2x + 5$. *What does the slope represent?*

(A) The starting value

(B) The rate of change per unit increase in x

(C) The x-intercept

(D) The total value of y

Find more at
ViewMath.com/IN-Grade8

ViewMath.com

28. A two-way table has 3 rows and 4 columns (not counting totals). How many joint frequencies are there?

(A) 3

(B) 4

(C) 7

(D) 12

29. A tree diagram shows flipping a coin and rolling a die (1–6). How many total outcomes are in the sample space?

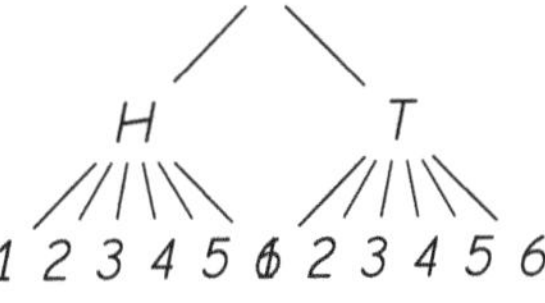

(A) 6

(B) 8

(C) 12

(D) 36

30. The diagram shows all arrangements of the letters A, B, C. How many permutations are there?

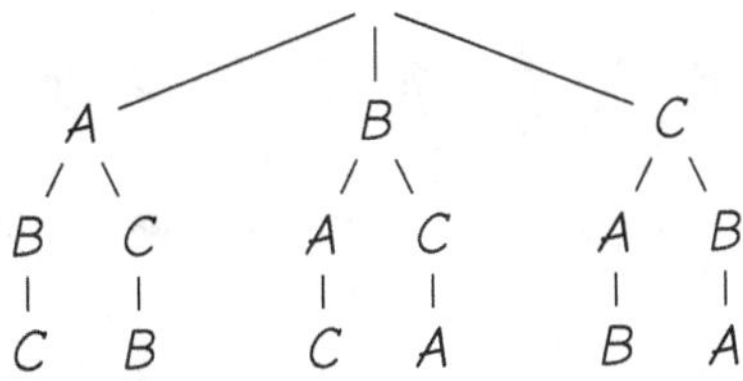

(A) 3

(B) 6

(C) 9

(D) 12

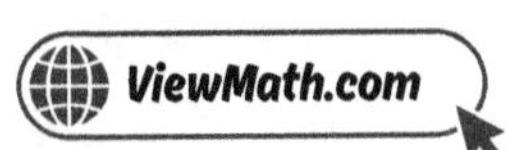

⭐ End of Practice Test 2 ⭐

Great job finishing the test!

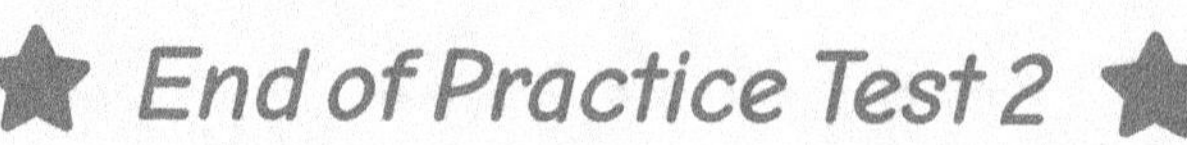

☑ My Score

I got ______________ out of 30 questions right.

*Check your answers in the **Answer Key** at the back of the book.*

💡 *Review any questions you missed. That's how we learn!*

📊 Check Your Score Online!

Visit **ViewMath Academy** to enter your answers and see which topics you need to review. You can also explore lessons, take quizzes, track your scores, and save your progress!

viewmath.com/score/8.1.IN.17

Or go to viewmath.com/score and enter code: 8.1.IN.17

Practice Test 3

30 Questions

✏️ Before You Start ✏️

- ✔ **Read each question carefully** before choosing your answer.

- ✔ **Show your work** on scratch paper when you need to.

- ✔ **Skip hard questions** and come back to them later.

- ✔ **Check your answers** when you're done.

- ✔ **Take your time** — there's no rush!

⭐ You've Got This! ⭐

Do your best and show what you know!

1. Which number below can be written as a fraction $\frac{a}{b}$ where a and b are integers and $b \neq 0$?

   (A) $\sqrt{3}$

   (B) π

   (C) $0.\overline{81}$

   (D) $\sqrt{11}$

2. Which is the best approximation of $\sqrt{30}$ to one decimal place?

   (A) 5.1

   (B) 5.3

   (C) 5.5

   (D) 5.8

3. Which is larger: $4\sqrt{2}$ or $\sqrt{30}$?

   (A) $4\sqrt{2}$, because $4 > \sqrt{30}$

   (B) $\sqrt{30}$, because $30 > 2$

   (C) $4\sqrt{2}$, because $4\sqrt{2} \approx 5.66 > 5.48 \approx \sqrt{30}$

   (D) They are equal.

4. Which of the following is equal to $\frac{8^6}{8^6}$?

   (A) 8^{36}

   (B) 8^0

   (C) 0

   (D) 8

5. A square has an area of 196 square centimeters. What is the length of one side?

   (A) $13\ cm$

   (B) $49\ cm$

   (C) $14\ cm$

   (D) $98\ cm$

Find more at
ViewMath.com/IN-Grade8

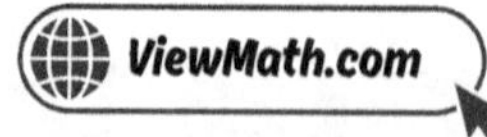

6. What is $\frac{8 \times 10^7}{2 \times 10^3}$?

- (A) 4×10^4
- (B) 6×10^4
- (C) 4×10^{10}
- (D) 16×10^{10}

7. Machine A fills $y = 12x$ bottles per hour. Machine B fills 50 bottles in 5 hours. Which machine is faster?

- (A) Machine A
- (B) Machine B
- (C) They fill at the same rate.
- (D) Cannot be determined.

8. A line passes through $(2, 7)$ and $(5, 1)$. What is the slope?

- (A) -2
- (B) 2
- (C) -3
- (D) 3

9. Which equation has the steepest line?

- (A) $y = x + 1$
- (B) $y = 3x - 2$
- (C) $y = -5x + 4$
- (D) $y = \frac{1}{2}x + 6$

10. Two trains leave the same station heading in opposite directions. Train A goes 60 mph and Train B goes 80 mph. After how many hours are they 420 miles apart?

- (A) 2
- (B) 2.5
- (C) 3
- (D) 3.5

11. A function is defined by the table below.

x	0	1	2	3
$f(x)$	-1	3	7	11

What is $f(0) + f(2)$?

(A) 2

(B) 6

(C) 8

(D) 10

12. Function A: $y = -x + 10$. Function B: $y = -4x + 10$. Which function decreases faster? Write A or B.

Your Answer:

13. A table shows $(0, 3)$, $(1, 7)$, $(2, 11)$, $(3, 15)$. Is this linear or nonlinear?

(A) Nonlinear, because the outputs are odd numbers only.

(B) Linear, because the change in y is always 4.

(C) Nonlinear, because the outputs get larger.

(D) Linear, because the first output is positive.

14. A line passes through $(0, 4)$ and $(5, 24)$. Write its equation.

Your Answer:

15. A graph of water in a tank is a straight line sloping downward. What is happening?

(A) Water is being added at a constant rate.

(B) Water is leaking out at a constant rate.

(C) The tank is full and not changing.

(D) Water is being added at an increasing rate.

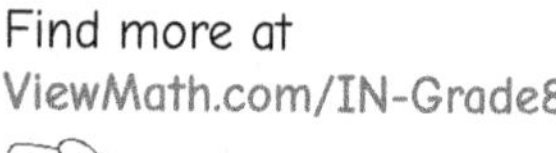

Find more at
ViewMath.com/IN-Grade8

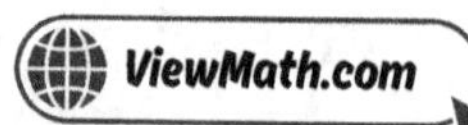

16. Point $R(-1, 4)$ is rotated 90° counterclockwise around the origin. What are the coordinates of R'?

(A) $(4, 1)$

(B) $(-4, -1)$

(C) $(1, -4)$

(D) $(-4, 1)$

17. Two quadrilaterals are shown below. Which statement is correct?

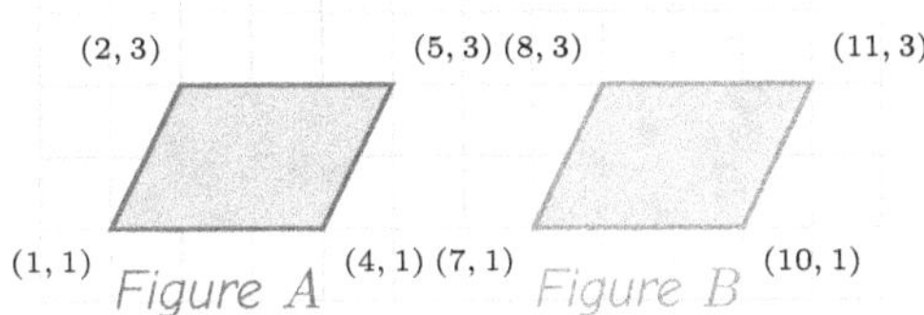

(A) They are congruent — mapped by a translation of 6 units right.

(B) They are not congruent — different side lengths.

(C) They are similar but not congruent.

(D) They are congruent — mapped by a reflection.

18. Point $A(0, -5)$ is rotated 90° counterclockwise around the origin. Where does A' land?

(A) $(5, 0)$

(B) $(0, 5)$

(C) $(-5, 0)$

(D) $(0, -5)$

19. Two similar triangles have a scale factor of 4 from the smaller to the larger. If a side of the smaller triangle is 3 cm, what is the corresponding side of the larger triangle?

(A) 7 cm

(B) 0.75 cm

(C) 12 cm

(D) 9 cm

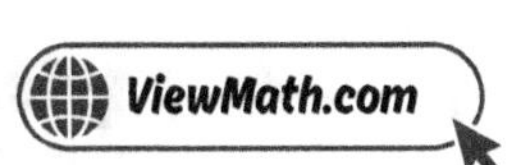

20. In an isosceles triangle, the two base angles are each 54°. What is the vertex angle?

(A) 54°

(B) 72°

(C) 108°

(D) 126°

21. A baseball diamond is a square with 90 ft sides. How far is it from home plate to second base (the diagonal)?

(A) 180 ft

(B) $90\sqrt{2}$ ft

(C) 45 ft

(D) $\sqrt{90}$ ft

22. Find the distance between points A and B shown on the coordinate plane.

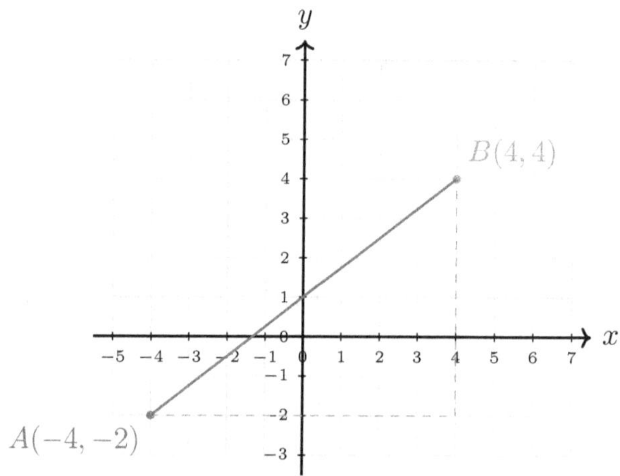

Your Answer:

23. The volume of a sphere with radius r is $\frac{4}{3}\pi r^3$. If the radius is tripled, the new volume is:

(A) 3 times the original

(B) 9 times the original

(C) 27 times the original

(D) 81 times the original

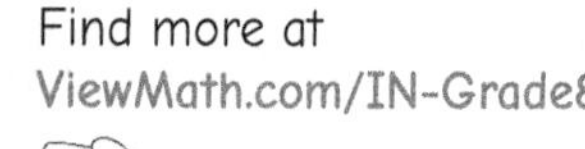

Find more at
ViewMath.com/IN-Grade8

 ViewMath.com

24. A square pyramid has base side 7 cm and height 12 cm. What is the volume?

(A) $196\ cm^3$

(B) $588\ cm^3$

(C) $294\ cm^3$

(D) $98\ cm^3$

25. A scatter plot has a strong upward trend. Which r-value (correlation) is most likely?

(A) $r = -0.9$

(B) $r = 0.1$

(C) $r = 0.9$

(D) $r = 0$

26. Data: $(1, 2), (2, 4), (3, 7), (4, 8), (5, 11)$. Estimate the slope of a reasonable trend line.

Your Answer:

27. A positive residual means:

(A) The predicted value is higher than the actual value.

(B) The actual value is higher than the predicted value.

(C) The slope of the model is positive.

(D) The data point is an outlier.

28. A survey asks students about their favorite sport (soccer or basketball) and grade (7th or 8th). 20 7th-graders chose soccer. This 20 is a:

(A) marginal frequency

(B) joint frequency

(C) relative frequency

(D) total frequency

29. Two events are independent if:

(A) They cannot happen at the same time

(B) The outcome of one does not affect the other

(C) They always happen together

(D) They share the same sample space

Find more at
ViewMath.com/IN-Grade8

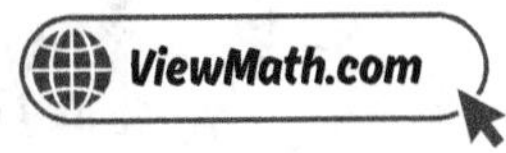

30. How many 3-letter codes can be made from the letters A, B, C, D if no letter is repeated?

(A) 12

(B) 24

(C) 64

(D) 4

 End of Practice Test 3 

Great job finishing the test!

📋 My Score

I got _____________ out of 30 questions right.

*Check your answers in the **Answer Key** at the back of the book.*

💡 *Review any questions you missed. That's how we learn!*

📊 Check Your Score Online!

Visit **ViewMath Academy** to enter your answers and see which topics you need to review. You can also explore lessons, take quizzes, track your scores, and save your progress!

viewmath.com/score/8.1.IN.18

Or go to *viewmath.com/score* and enter code: 8.1.IN.18

Practice Test 4

☑ 30 Questions

✏ Before You Start ✏

- ✔ **Read each question carefully** before choosing your answer.
- ✔ **Show your work** on scratch paper when you need to.
- ✔ **Skip hard questions** and come back to them later.
- ✔ **Check your answers** when you're done.
- ✔ **Take your time** — there's no rush!

⭐ You've Got This! ⭐

Do your best and show what you know!

1. Which of the following numbers is irrational?

(A) $\frac{5}{8}$

(B) $0.\overline{6}$

(C) $\sqrt{9}$

(D) $\sqrt{7}$

2. Approximate $\sqrt{68}$ to one decimal place.

Your Answer

3. A student claims that $2\sqrt{5} = \sqrt{10}$. Is this correct?

(A) Yes, because $2 \times 5 = 10$.

(B) No, $2\sqrt{5} = \sqrt{20}$, not $\sqrt{10}$.

(C) Yes, multiplication distributes into the square root.

(D) No, $2\sqrt{5} = 4\sqrt{5}$.

4. Evaluate $5^0 + 5^{-1}$. Express your answer as a decimal.

Your Answer

5. What is $\sqrt{49}$?

(A) 6

(B) 7

(C) 8

(D) 24.5

6. What is $5.2 \times 10^6 + 3.8 \times 10^6$?

(A) 9×10^6

(B) 9×10^{12}

(C) 19.76×10^6

(D) 9×10^{36}

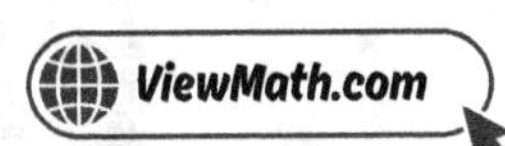

7. Which ordered pair could NOT lie on the graph of a proportional relationship?

(A) $(0,0)$

(B) $(1,5)$

(C) $(3,15)$

(D) $(2,13)$

8. A line passes through $(0,4)$ and $(8,4)$. What is its slope?

Your Answer

9. Write the equation of a line with slope $\frac{2}{3}$ that passes through $(0,-4)$.

(A) $y = \frac{2}{3}x + 4$

(B) $y = \frac{2}{3}x - 4$

(C) $y = -4x + \frac{2}{3}$

(D) $y = 4x + \frac{2}{3}$

10. Two numbers add to 25 and differ by 7. What is the larger number?

(A) 14

(B) 15

(C) 16

(D) 17

11. If $h(x) = 2x^2$, what is $h(5)$?

(A) 20

(B) 25

(C) 50

(D) 100

12. Function A: $y = 2x + 15$. Function B: $y = 5x + 6$. What is the value of each function at $x = 3$?

Your Answer:

13. The table below shows a function. Is it linear or nonlinear?

x	0	1	2	3	4
y	1	2	5	10	17

(A) Linear, because y always increases.

(B) Linear, because x increases by 1 each time.

(C) Nonlinear, because the differences in y are $1, 3, 5, 7$ — not constant.

(D) Nonlinear, because the outputs are all positive.

14. Find the equation of the line through $(1, 3)$ and $(4, 12)$.

Your Answer

15. A bathtub fills up, then someone soaks, then the water drains. Which describes the water level graph?

(A) Decreasing, constant, increasing

(B) Increasing, decreasing, constant

(C) Increasing, constant, decreasing

(D) Constant, increasing, decreasing

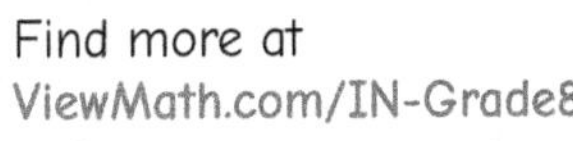

Find more at
ViewMath.com/IN-Grade8

ViewMath.com

16. Look at the coordinate plane. Which transformation maps triangle PQR to triangle $P'Q'R'$?

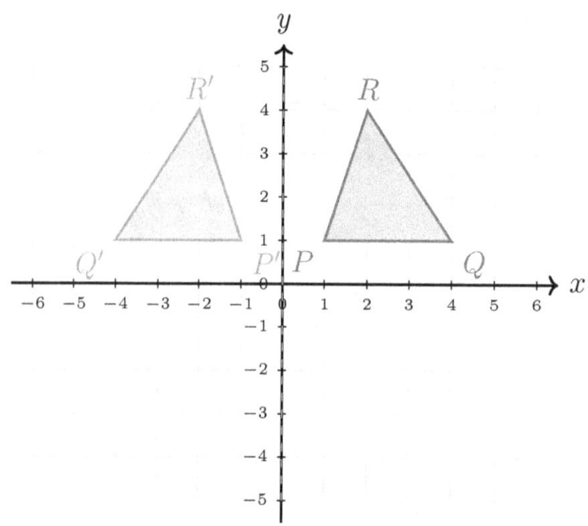

(A) Translation 5 units left

(B) Reflection over the y-axis

(C) Rotation 90° clockwise

(D) Reflection over the x-axis

17. Rectangle $ABCD$ has $AB = 6$ and $BC = 10$. Rectangle $WXYZ$ has $WX = 10$ and $XY = 6$. Are they congruent?

(A) No, because the side lengths are in different order.

(B) No, because $6 \neq 10$.

(C) Yes, because the sets of side lengths are the same.

(D) Cannot be determined.

18. A rectangle has vertex $(-6, 2)$. After a translation of $(4, -5)$, what is the new position of this vertex?

(A) $(-10, 7)$

(B) $(-2, -3)$

(C) $(2, -3)$

(D) $(-2, 3)$

19. *If two figures are congruent, are they also similar?*

    (A) *Yes, with scale factor $k = 1$.*

    (B) *No, congruent and similar are different properties.*

    (C) *Only if they are triangles.*

    (D) *Only if they have the same orientation.*

20. *An exterior angle of a triangle is $140°$. What is the adjacent interior angle?*

    (A) $40°$

    (B) $140°$

    (C) $50°$

    (D) $220°$

21. *Is a triangle with sides 7, 24, 25 a right triangle?*

    (A) *Yes, because $7 + 24 > 25$.*

    (B) *Yes, because $7^2 + 24^2 = 25^2$.*

    (C) *No, because $7^2 + 24^2 \neq 25^2$.*

    (D) *No, because all sides must be equal.*

22. *What is the distance between $(-1, 2)$ and $(2, 6)$?*

    (A) 25

    (B) 7

    (C) 5

    (D) $\sqrt{13}$

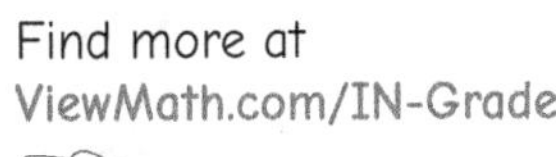
Find more at
ViewMath.com/IN-Grade8

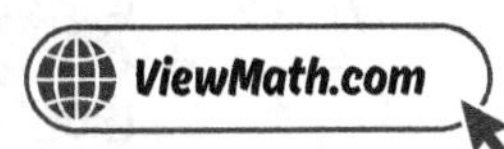

23. What is the volume of the cone shown below in terms of π?

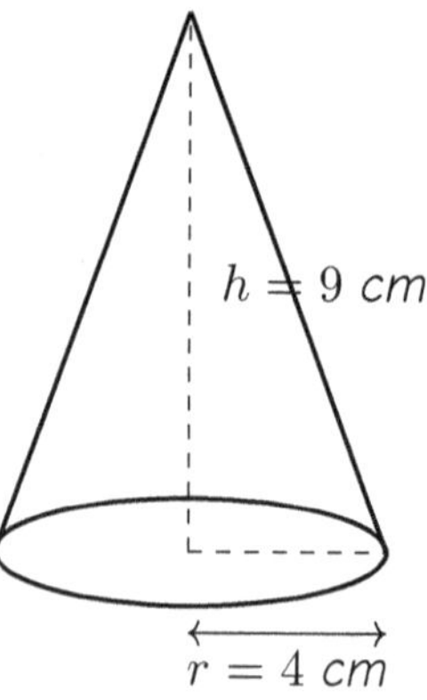

(A) $144\pi\ cm^3$

(B) $48\pi\ cm^3$

(C) $36\pi\ cm^3$

(D) $108\pi\ cm^3$

24. A pyramid has a pentagonal base with area 35 cm^2 and height 6 cm. What is the volume?

Your Answer:

25. A scatter plot of shoe size versus favorite color shows dots scattered randomly with no pattern. What type of association is this?

(A) Positive

(B) Negative

(C) Linear

(D) No association

26. A trend line passes through $(3, 7)$ and $(9, 19)$. What is the slope?

(A) 1

(B) 2

(C) 3

(D) 4

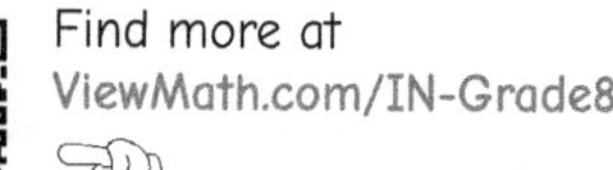
Find more at
ViewMath.com/IN-Grade8

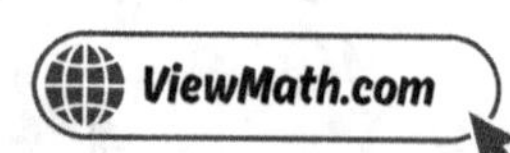

27. In the model $y = 2.5x + 1$, the actual value at $x = 4$ is $y = 12$. What is the residual?

(A) 0

(B) 1

(C) -1

(D) 11

28.

	Pass	Fail	Total
Studied	30	5	35
Did not study	10	15	25
Total	40	20	60

What fraction of all students passed?

(A) $\frac{30}{60}$

(B) $\frac{40}{60}$

(C) $\frac{35}{60}$

(D) $\frac{20}{60}$

29. A bag has 4 red and 6 blue marbles. Draw a tree diagram for picking 2 marbles without replacement. Find $P(\text{both blue})$.

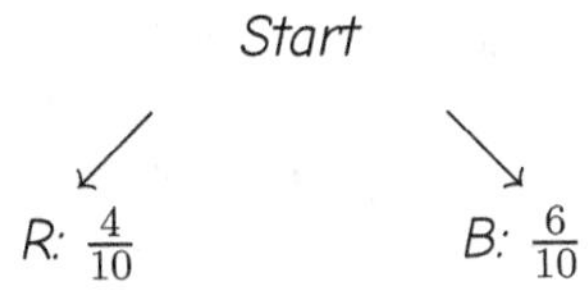

Your Answer

30. A class of 10 students needs to select 2 for a project. How many combinations?

(A) 20

(B) 45

(C) 90

(D) 100

 End of Practice Test 4 

Great job finishing the test!

✅ My Score

I got _____________ out of 30 questions right.

*Check your answers in the **Answer Key** at the back of the book.*

💡 *Review any questions you missed. That's how we learn!*

📊 Check Your Score Online!

Visit **ViewMath Academy** to enter your answers and see which topics you need to review. You can also explore lessons, take quizzes, track your scores, and save your progress!

viewmath.com/score/8.1.IN.19

Or go to viewmath.com/score and enter code: 8.1.IN.19

Practice Test 5

30 Questions

✏️ Before You Start ✏️

✔ **Read each question carefully** before choosing your answer.

✔ **Show your work** on scratch paper when you need to.

✔ **Skip hard questions** and come back to them later.

✔ **Check your answers** when you're done.

✔ **Take your time** — there's no rush!

⭐ You've Got This! ⭐

Do your best and show what you know!

1. A calculator shows $\sqrt{144} = 12$. Is $\sqrt{144}$ rational or irrational? Explain.

   Your Answer:

2. Between which two consecutive integers does $\sqrt{50}$ lie?

   (A) 6 and 7

   (B) 7 and 8

   (C) 8 and 9

   (D) 24 and 26

3. If the side length of a square is $\sqrt{50}$ cm, what is the perimeter?

   (A) $4\sqrt{50} \approx 28.3$ cm

   (B) 50 cm

   (C) $\sqrt{200} \approx 14.1$ cm

   (D) 200 cm

4. Study the number line below. Which expression corresponds to the point marked with a star (★)?

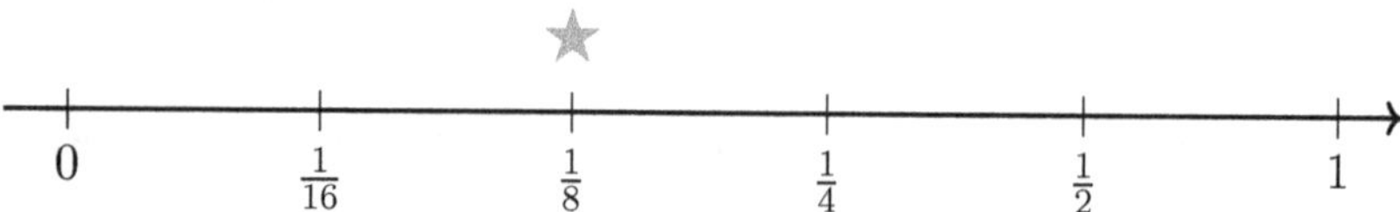

   (A) 2^{-2}

   (B) 2^{-3}

   (C) 2^{-4}

   (D) 2^{-1}

5. Which of the following is a perfect square?

   (A) 50

   (B) 72

   (C) 81

   (D) 90

Find more at
ViewMath.com/IN-Grade8

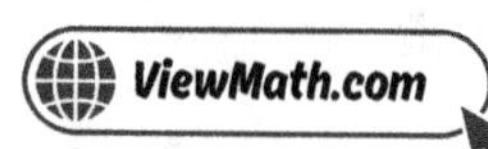

6. Compute $(8 \times 10^{-4})(5 \times 10^{-3})$. Write your answer in scientific notation.

Your Answer:

7. Store A sells apples for \$2 per pound. Store B's prices are shown in the table.

Pounds	3	6	9
Cost (\$)	7.50	15.00	22.50

Which store has the lower unit price?

(A) Store A

(B) Store B

(C) They charge the same price.

(D) Not enough information.

8. A bathtub drains from 40 gallons to 10 gallons in 15 minutes. What is the rate of change?

(A) -2 gal/min

(B) 2 gal/min

(C) -30 gal/min

(D) 30 gal/min

9. A line passes through $(0, -2)$ and $(5, 8)$. What is its equation?

(A) $y = 2x + 2$

(B) $y = -2x + 2$

(C) $y = 2x - 2$

(D) $y = -2x - 2$

10. A rectangle has a perimeter of 48 m. Its length is twice its width. Find the dimensions.

Your Answer:

Find more at
ViewMath.com/IN-Grade8

ViewMath.com

11. If $f(x) = 2x + 9$, what is $f(5)$?

Your Answer:

12. Function P is given by the table:

x	0	1	2	3
y	4	7	10	13

Function Q: $y = 2x + 5$. Which function has a greater rate of change?

(A) Function P

(B) Function Q

(C) They have the same rate of change.

(D) Cannot be determined.

13. A table shows $(1, 1)$, $(2, 8)$, $(3, 27)$, $(4, 64)$. Is this linear or nonlinear? What pattern do you see?

Your Answer:

14. A gym charges a one-time fee of \$60 plus \$25 per month. What is the total cost after 8 months?

(A) \$200

(B) \$260

(C) \$285

(D) \$480

15. A ball is thrown straight up. Its height over time:

(A) Increases, then decreases — the graph is non-linear.

(B) Increases, then decreases — the graph is linear.

(C) Decreases only — the graph is linear.

(D) Increases only — the graph is nonlinear.

Find more at
ViewMath.com/IN-Grade8

16. Point $N(0, 4)$ is rotated $180°$ around the origin. What are the coordinates of N'?

Your Answer:

17. $\triangle ABC \cong \triangle DEF$. If $\angle A = 55°$ and $\angle B = 80°$, what is $\angle F$?

(A) $55°$

(B) $80°$

(C) $45°$

(D) $135°$

18. What is the image of $(-4, 2)$ after a reflection over the $y-$axis?

(A) $(-4, -2)$

(B) $(4, -2)$

(C) $(4, 2)$

(D) $(2, -4)$

19. Are all squares similar to each other?

(A) No, because they can have different side lengths.

(B) No, because they can have different angles.

(C) Yes, because all squares have equal angles and proportional sides.

(D) Yes, but only if they have the same perimeter.

20. Corresponding angles formed by a transversal and two parallel lines are located:

(A) On opposite sides of the transversal, between the parallel lines

(B) On the same side of the transversal, in matching positions

(C) On opposite sides of the transversal, outside the parallel lines

(D) Adjacent to each other at one intersection

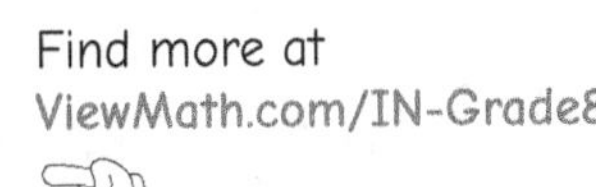

Find more at
ViewMath.com/IN-Grade8

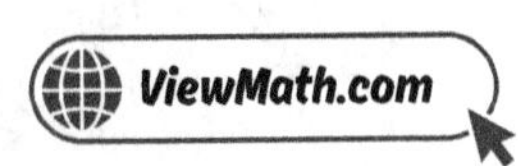

21. A right triangle has legs 10 and 24. What is the hypotenuse?

Your Answer:

22. What is the distance between $(-2, 1)$ and $(4, 9)$?

(A) 10

(B) $\sqrt{10}$

(C) 14

(D) 8

23. A traffic cone has radius 8 in and height 24 in. What is its volume? Use $\pi \approx 3.14$.

(A) 4,823.0 in^3

(B) 1,607.7 in^3

(C) 803.8 in^3

(D) 602.9 in^3

24. What is the volume of a square pyramid with base side 6 cm and height 10 cm?

(A) 360 cm^3

(B) 120 cm^3

(C) 180 cm^3

(D) 60 cm^3

25. Which scatter plot pattern suggests a linear association?

(A) Dots that form a U-shape

(B) Dots scattered randomly

(C) Dots that roughly follow a straight line

(D) Dots that form a circle

26. Which describes a "good fit"?

(A) All data points are far from the line.

(B) Data points are close to the line.

(C) All data points are below the line.

(D) The line is vertical.

Find more at
ViewMath.com/IN-Grade8

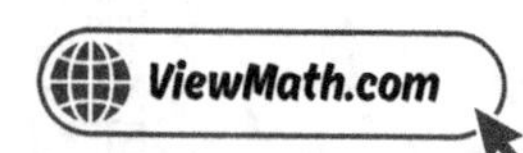

27. A trend line gives $y = -3x + 45$. Data was collected for $x = 1$ to $x = 10$. Predict y when $x = 5$ and when $x = 30$. State which prediction is more reliable and why.

Your Answer:

28. A relative frequency table shows the value 0.15 in one cell. If the total is 200, what is the frequency?

(A) 15

(B) 30

(C) 10

(D) 20

29. Draw a tree diagram for flipping a coin twice. List all outcomes and find P(at least one head).

Your Answer:

30. You have 6 toppings and want to choose 3 for a pizza (order doesn't matter). How many ways?

(A) 120

(B) 20

(C) 18

(D) 216

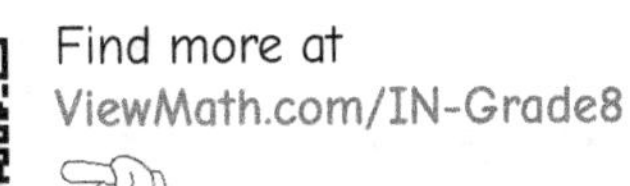

 # End of Practice Test 5 

Great job finishing the test!

 **My Score**

I got ___________ out of 30 questions right.

*Check your answers in the **Answer Key** at the back of the book.*

💡 *Review any questions you missed. That's how we learn!*

📊 Check Your Score Online!

Visit **ViewMath Academy** to enter your answers and see which topics you need to review. You can also explore lessons, take quizzes, track your scores, and save your progress!

viewmath.com/score/8.1.IN.20

Or go to viewmath.com/score and enter code: 8.1.IN.20

6

Practice Test 6

📋 30 Questions

✏️ Before You Start ✏️

- ✔ **Read each question carefully** before choosing your answer.
- ✔ **Show your work** on scratch paper when you need to.
- ✔ **Skip hard questions** and come back to them later.
- ✔ **Check your answers** when you're done.
- ✔ **Take your time** — there's no rush!

⭐ You've Got This! ⭐

Do your best and show what you know!

1. Is $\frac{5}{6}$ rational or irrational? Explain.

2. A student says $\sqrt{45}$ is between 6 and 7. Is the student correct?

(A) Yes, because $6^2 = 36$ and $7^2 = 49$.

(B) No, it is between 5 and 6.

(C) No, it is between 7 and 8.

(D) No, $\sqrt{45} = 9$ exactly.

3. Order from least to greatest: π, $\sqrt{10}$, 3.2.

(A) π, 3.2, $\sqrt{10}$

(B) 3.2, π, $\sqrt{10}$

(C) $\sqrt{10}$, π, 3.2

(D) π, $\sqrt{10}$, 3.2

4. What is the value of $(-2)^0$?

(A) -2

(B) 0

(C) 1

(D) -1

5. A cube has a volume of 343 cubic inches. What is the length of one edge?

(A) 7 in.

(B) 49 in.

(C) $114.\overline{3}$ in.

(D) 17 in.

6. What is $7.5 \times 10^8 - 2.5 \times 10^8$?

(A) 5×10^0

(B) 5×10^8

(C) 10×10^8

(D) 5×10^{16}

Find more at
ViewMath.com/IN-Grade8

7. A line passes through the origin and the point $(6, 18)$. What is the equation of the line?

(A) $y = 6x$

(B) $y = 18x$

(C) $y = 3x$

(D) $y = \frac{1}{3}x$

8. A car's odometer reads 120 miles at 2:00 PM and 270 miles at 5:00 PM. What is the car's speed (rate of change)?

(A) 45 mph

(B) 50 mph

(C) 60 mph

(D) 90 mph

9. Which equation represents a line with slope -2 and y-intercept 7?

(A) $y = 7x - 2$

(B) $y = -2x + 7$

(C) $y = 2x + 7$

(D) $y = -2x - 7$

10. Maria has 15 coins, all nickels and dimes. The coins are worth \$1.10. How many dimes does she have?

(A) 5

(B) 6

(C) 7

(D) 8

11. If $f(x) = 7 - x$, what is $f(10)$?

(A) -3

(B) 3

(C) 17

(D) -17

Find more at
ViewMath.com/IN-Grade8

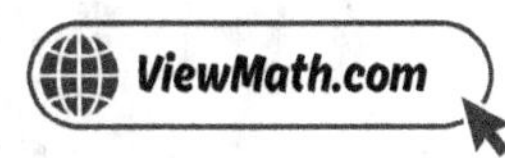

12. The graph below shows Function A and Function B. Which function has a greater rate of change?

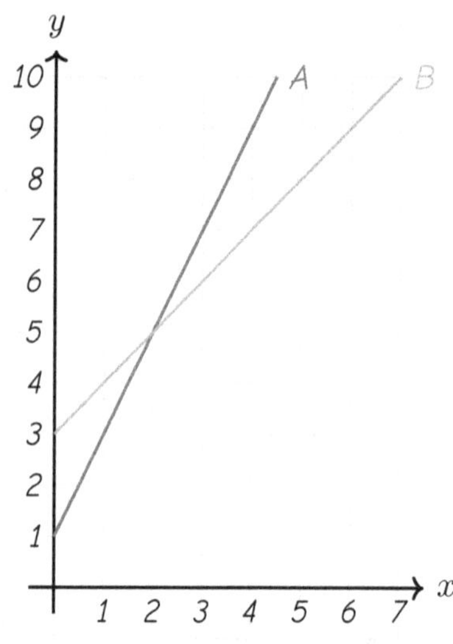

(A) Function A

(B) Function B

(C) They have the same rate of change.

(D) Cannot be determined from the graph.

13. A function increases by 6 for every increase of 1 in x. Is it linear or nonlinear?

Your Answer:

14. A candle is 12 inches tall and burns at 2 inches per hour. What is the function for the candle's height y after x hours?

(A) $y = 2x + 12$

(B) $y = -2x + 12$

(C) $y = 12x - 2$

(D) $y = -12x + 2$

15. A graph shows temperature over a day. From 6 AM to noon, the graph goes up steeply. From noon to 3 PM, t goes up slowly. What can you conclude?

(A) The temperature decreased between noon and 3 PM.

(B) The temperature increased faster in the morning than in the afternoon.

(C) The temperature was constant from noon to 3 PM.

(D) The temperature increased at the same rate all day.

16. A student says that a reflection changes the area of a figure. Is this correct?

(A) Yes, because the figure flips.

(B) Yes, because one dimension reverses.

(C) No, because reflections preserve all measurements.

(D) No, but reflections change the perimeter.

17. Two congruent rectangles each have a perimeter of 28 cm. One has a length of 9 cm. What is its width?

Your Answer:

18. A point (a, b) is translated by $(5, -1)$ to get image $(2, 7)$. What is (a, b)?

Your Answer:

19. Which statement about similar figures is always true?

(A) They have equal side lengths.

(B) They have equal angle measures.

(C) They have the same perimeter.

(D) They have the same area.

20. A triangle has angles $90°$, $x°$, and $(x + 10)°$. Find x.

Your Answer:

Find more at
ViewMath.com/IN-Grade8

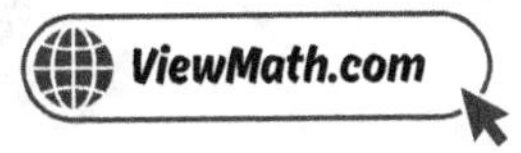

21. A right triangle has legs 1 and 1. What is the exact length of the hypotenuse?

(A) 2

(B) $\sqrt{2}$

(C) 1

(D) $\sqrt{3}$

22. Find the distance between $(0,0)$ and $(2,3)$. Leave your answer in simplest radical form.

Your Answer

23. A cone has radius 3 m and height 12 m. What is its volume in terms of π?

(A) $108\pi\ m^3$

(B) $36\pi\ m^3$

(C) $12\pi\ m^3$

(D) $324\pi\ m^3$

24. A triangular pyramid has a base triangle with area 30 cm^2 and a height of 9 cm. What is the volume?

(A) $270\ cm^3$

(B) $90\ cm^3$

(C) $135\ cm^3$

(D) $45\ cm^3$

25. Which variable typically goes on the x-axis in a scatter plot?

(A) The dependent variable

(B) The response variable

(C) The independent (explanatory) variable

(D) The variable with the largest values

26. A line passes through $(0,3)$ and $(4,11)$. What is the y-intercept?

(A) 0

(B) 4

(C) 3

(D) 11

Find more at
ViewMath.com/IN-Grade8

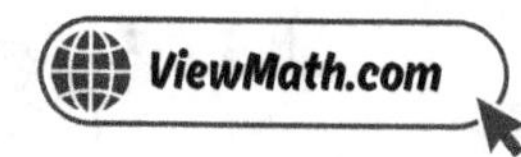

27. Which statement about slope in a linear model is true?

(A) A positive slope means y decreases as x increases.

(B) A negative slope means y increases as x increases.

(C) A slope of zero means y stays constant.

(D) The slope is always positive in real data.

28. What is a relative frequency?

(A) The total number of data values

(B) A frequency expressed as a fraction or percentage of a total

(C) The difference between two frequencies

(D) The mode of the data

29. A bag has 3 red, 5 blue, and 2 green marbles. What is $P(\text{blue})$?

(A) $\frac{1}{5}$

(B) $\frac{5}{10}$

(C) $\frac{3}{10}$

(D) $\frac{2}{5}$

30. The Fundamental Counting Principle states that if event A has m outcomes and event B has n outcomes, the total number of outcomes for A followed by B is:

(A) $m + n$

(B) $m \times n$

(C) $m - n$

(D) m^n

Find more at
ViewMath.com/IN-Grade8

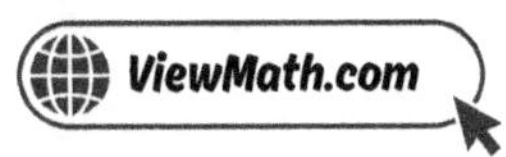

 # End of Practice Test 6 

Great job finishing the test!

 **My Score**

I got _____________ out of 30 questions right.

*Check your answers in the **Answer Key** at the back of the book.*

💡 *Review any questions you missed. That's how we learn!*

📊 **Check Your Score Online!**

Visit **ViewMath Academy** to enter your answers and see which topics you need to review. You can also explore lessons, take quizzes, track your scores, and save your progress!

viewmath.com/score/8.1.IN.21

Or go to viewmath.com/score and enter code: 8.1.IN.21

Practice Test 7

 30 Questions

✏️ Before You Start ✏️

- ✔ **Read each question carefully** before choosing your answer.
- ✔ **Show your work** on scratch paper when you need to.
- ✔ **Skip hard questions** and come back to them later.
- ✔ **Check your answers** when you're done.
- ✔ **Take your time** — there's no rush!

 ⭐ You've Got This! ⭐ 

Do your best and show what you know!

1. Give an example of an irrational number between 1 and 2.

   Your Answer

2. Which list is in order from least to greatest?

   (A) $\sqrt{5}$, 2, $\frac{5}{2}$

   (B) 2, $\sqrt{5}$, $\frac{5}{2}$

   (C) $\frac{5}{2}$, 2, $\sqrt{5}$

   (D) 2, $\frac{5}{2}$, $\sqrt{5}$

3. Estimate $\sqrt{3} + \sqrt{7}$ to one decimal place.

   Your Answer

4. Which expression is equivalent to $\frac{9^3 \cdot 9^2}{9^4}$?

   (A) 9^{-1}

   (B) 9^0

   (C) 9^1

   (D) 9^{24}

5. A cube-shaped box has a volume of 216 cubic centimeters. What is the edge length of the box?

   Your Answer

6. What is $(3 \times 10^4)(2 \times 10^5)$?

   (A) 6×10^9

   (B) 6×10^{20}

   (C) 5×10^9

   (D) 6×10^1

7. A painter paints 3 rooms in 6 hours. At this rate, how many hours will it take to paint 8 rooms?

Your Answer:

8. Which pair of points gives a slope of $\frac{1}{2}$?

(A) $(0,0)$ and $(2,4)$

(B) $(1,3)$ and $(5,5)$

(C) $(2,1)$ and $(4,3)$

(D) $(0,1)$ and $(1,4)$

9. What is the slope and y-intercept of $y = 5x - 3$?

(A) slope $= -3$, y-intercept $= 5$

(B) slope $= 5$, y-intercept $= -3$

(C) slope $= 3$, y-intercept $= 5$

(D) slope $= 5$, y-intercept $= 3$

10. A rectangle's perimeter is 34 cm. The length is 5 cm more than the width. What is the width?

(A) 4 cm

(B) 5 cm

(C) 6 cm

(D) 7 cm

11. If $f(x) = \frac{x+4}{2}$, what is $f(6)$?

Your Answer:

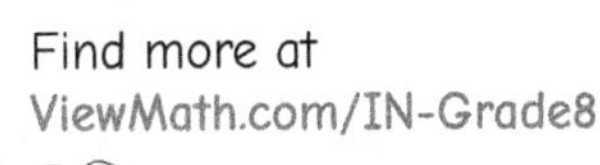
Find more at
ViewMath.com/IN-Grade8

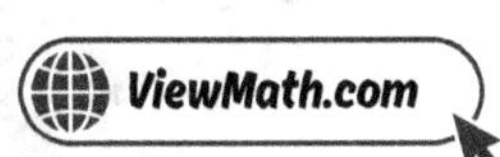
ViewMath.com

12. Function R from the previous table has a rate of change of:

x	0	2	4	6
y	1	9	17	25

(A) 2

(B) 3

(C) 4

(D) 8

13. The equation $A = s^2$ gives the area of a square with side length s. Is this function linear or nonlinear?

(A) Linear, because area is always positive.

(B) Linear, because it has only one variable.

(C) Nonlinear, because the variable is squared.

(D) Nonlinear, because area is measured in square units.

14. A table shows $(0, 8)$, $(3, 20)$, $(6, 32)$. What is the equation?

(A) $y = 3x + 8$

(B) $y = 4x + 8$

(C) $y = 8x + 4$

(D) $y = 6x + 8$

15. A graph of amount of gas in a car over time goes downward. Is the function increasing, decreasing, or constant?

Your Answer:

16. Which transformation flips a figure over a line?

(A) Translation

(B) Rotation

(C) Dilation

(D) Reflection

Find more at
ViewMath.com/IN-Grade8

ViewMath.com

17. Two figures are congruent if one can be mapped onto the other using which of the following?

(A) Only translations

(B) Only dilations

(C) A sequence of rigid transformations

(D) Any single transformation

18. A point is dilated by factor 3 from the origin. If the image is $(9, -15)$, what was the original point?

(A) $(27, -45)$

(B) $(3, -5)$

(C) $(6, -12)$

(D) $(12, -18)$

19. All circles are similar to each other. Why?

(A) They all have the same radius.

(B) One can always be dilated to match the other.

(C) They all have the same area.

(D) Circles cannot be dilated.

20. In a triangle, the angles are $(2x + 5)°$, $(3x)°$, and $(x + 25)°$. Find x.

Your Answer:

21. A right triangle has hypotenuse 17 and one leg 8. What is the other leg?

(A) 9

(B) 15

(C) 25

(D) $\sqrt{225}$

22. What is the distance between $(1, 1)$ and $(4, 5)$?

(A) 25

(B) 7

(C) $\sqrt{5}$

(D) 5

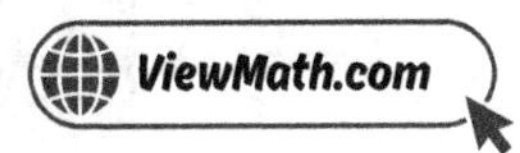

23. Find the volume of a cylinder with radius 6 cm and height 5 cm. Leave your answer in terms of π.

Your Answer:

24. A rectangular pyramid is shown. What is the volume?

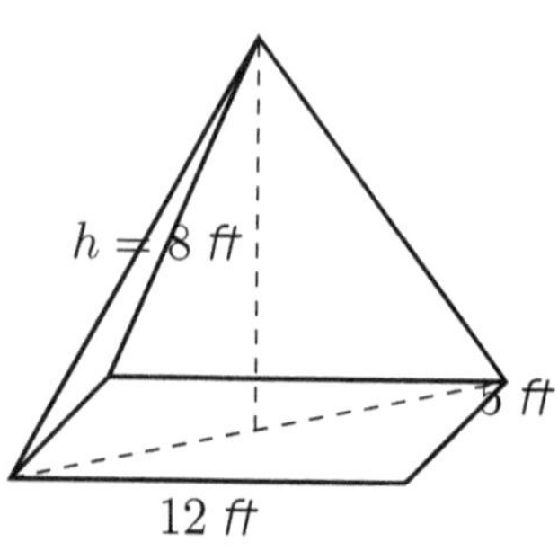

(A) $480\ ft^3$

(B) $160\ ft^3$

(C) $240\ ft^3$

(D) $320\ ft^3$

25. Does a strong association in a scatter plot prove that one variable causes the other? Explain.

Your Answer:

26. A scatter plot shows a strong nonlinear (curved) pattern. Should you draw a straight line of best fit?

(A) Yes — always draw a straight line.

(B) No — a straight line does not fit curved data well.

(C) Yes — but only if there are outliers.

(D) No — you should never draw any line.

27. A model for plant height is $y = 0.4x + 2$ where x is days and y is height in cm. Using the graph, find the predicted height at day 10, and explain what the slope and y-intercept mean.

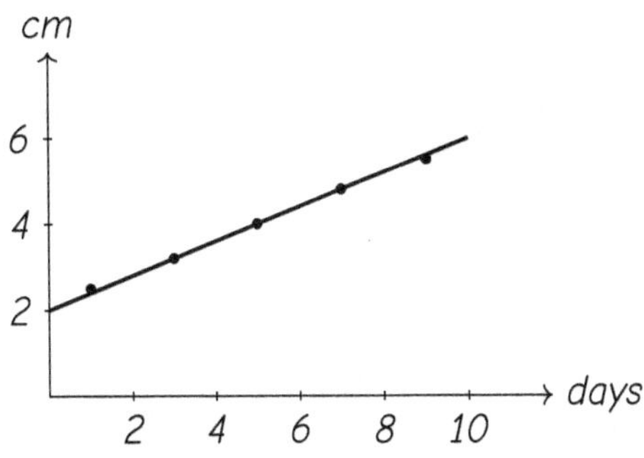

Your Answer:

28. What is a two-way frequency table?

(A) A table with two rows

(B) A table that shows frequencies for two categorical variables

(C) A table that shows only percentages

(D) A table with exactly two columns

29. $P(A) = 0.6$, $P(B) = 0.4$. If A and B are independent, $P(A \text{ and } B) = ?$

(A) 1.0

(B) 0.24

(C) 0.20

(D) 0.10

30. A coin is flipped 5 times. How many possible outcomes are there?

(A) 10

(B) 25

(C) 32

(D) 120

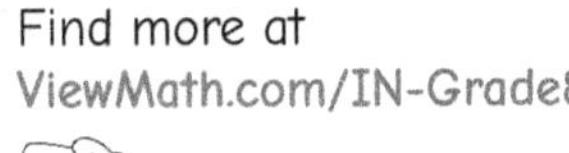
Find more at
ViewMath.com/IN-Grade8

 # End of Practice Test 7 

Great job finishing the test!

☑ My Score

I got _____________ out of 30 questions right.

*Check your answers in the **Answer Key** at the back of the book.*

💡 *Review any questions you missed. That's how we learn!*

📊 Check Your Score Online!

Visit **ViewMath Academy** to enter your answers and see which topics you need to review. You can also explore lessons, take quizzes, track your scores, and save your progress!

viewmath.com/score/8.1.IN.22

Or go to viewmath.com/score and enter code: 8.1.IN.22

Practice Test 8

☑ *30 Questions*

✏️ Before You Start ✏️

- ✔ **Read each question carefully** before choosing your answer.
- ✔ **Show your work** on scratch paper when you need to.
- ✔ **Skip hard questions** and come back to them later.
- ✔ **Check your answers** when you're done.
- ✔ **Take your time** — there's no rush!

⭐ *You've Got This!* ⭐

Do your best and show what you know!

1. Which of the following is a true statement?

   (A) Every square root is irrational.

   (B) Every integer is irrational.

   (C) Every integer is rational.

   (D) Every decimal is irrational.

2. The squares below have the given areas. Estimate the side length of each square to one decimal place.

   A   B   C

   Area = 18 sq units   Area = 40 sq units   Area = 72 sq units

   Your Answer:

3. A square patio has an area of 75 square feet. Estimate the side length to one decimal place.

   Your Answer:

4. Which shows $\frac{1}{2^4}$ written with a negative exponent?

   (A) $(-2)^4$

   (B) 2^{-4}

   (C) -2^4

   (D) 4^{-2}

5. Evaluate $\sqrt[3]{-125}$.

   Your Answer:

6. The bar chart compares the distances of four planets from the Sun. Which planet is approximately 10 times farther from the Sun than Planet A?

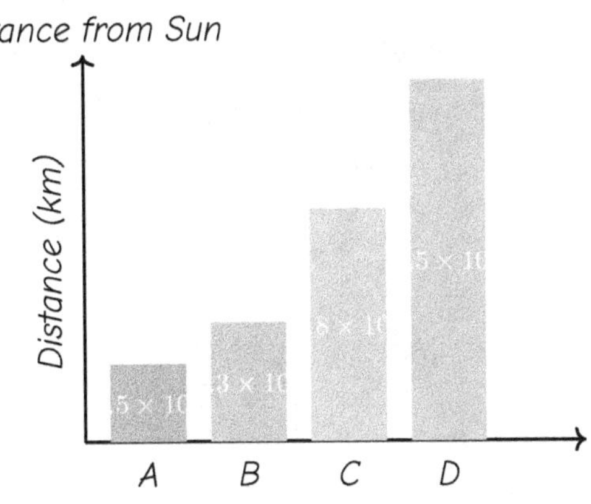

(A) Planet B

(B) Planet C

(C) Planet D

(D) None of the planets

7. Hose A fills a pool at $y = 15x$ gallons per minute. Hose B fills 100 gallons in 8 minutes. Which fills faster?

(A) Hose A

(B) Hose B

(C) They fill at the same rate.

(D) Not enough information.

8. A line has a slope of 0. Which of the following best describes the line?

(A) It is vertical.

(B) It goes up steeply from left to right.

(C) It is horizontal.

(D) It has undefined slope.

9. Using similar triangles, Mia draws a line and picks two different pairs of points. She gets slopes of $\frac{6}{3}$ and $\frac{10}{5}$. What can she conclude?

(A) She made a mistake — the slopes should be different.

(B) The slopes are 2 for both, confirming slope is constant on a line.

(C) The slopes are different because the triangles are different sizes.

(D) Similar triangles cannot be used to find slope.

10. One number is 4 more than another. Their sum is 36. Find both numbers.

Your Answer

11. The table below shows values for $f(x)$. If $f(x) = 15$, what is x?

x	0	1	2	3	4
$f(x)$	3	6	9	12	15

(A) 2

(B) 3

(C) 4

(D) 5

12. Function R is shown in the table:

x	0	2	4	6
y	1	9	17	25

Function S: $y = 3x + 1$. Which has a greater initial value?

(A) Function R

(B) Function S

(C) They have the same initial value.

(D) Cannot be determined.

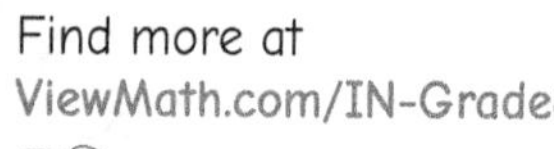

Find more at
ViewMath.com/IN-Grade8

ViewMath.com

13. Give an example of a nonlinear equation.

Your Answer:

14. The table shows a real-world function. Write the equation and explain what m and b represent.

Hours worked (x)	Pay (y)
0	$50
1	$62
2	$74
3	$86

Your Answer:

15. A student fills a glass with water, drinks half, then refills it. Describe the graph of water level vs. time.

Your Answer:

16. A point (a, b) is reflected over the x-axis, then reflected over the y-axis. What are the final coordinates?

Your Answer:

17. $\triangle PQR \cong \triangle STU$. If $\angle P = 72°$ and $\angle Q = 53°$, what is $\angle U$?

Your Answer:

Find more at
ViewMath.com/IN-Grade8

ViewMath.com

18. Which transformation maps (x, y) to $(-y, x)$?

(A) Reflection over the x-axis

(B) Reflection over the y-axis

(C) 90° counterclockwise rotation

(D) 180° rotation

19. Triangle A has sides 6, 8, 10. Triangle B has sides 9, 12, 15. What is the scale factor from A to B?

Your Answer

20. In the figure below, what is the value of y?

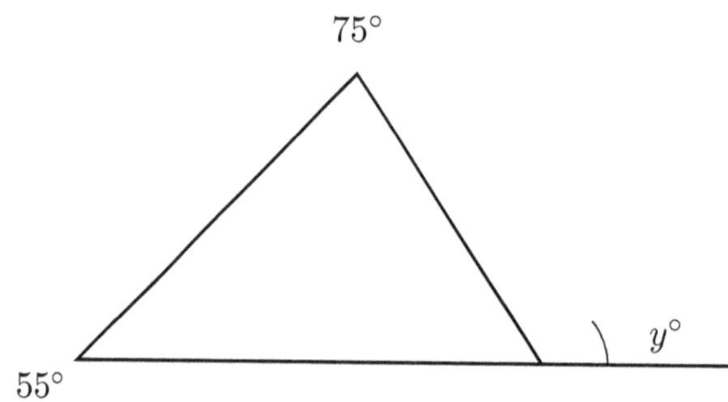

(A) 50°

(B) 105°

(C) 130°

(D) 125°

21. A rectangular room is 12 ft long and 5 ft wide. What is the diagonal distance across the floor?

Your Answer:

22. The distance formula is derived from:

(A) The area formula for a rectangle

(B) The Pythagorean Theorem

(C) The perimeter formula

(D) The midpoint formula

 Find more at
ViewMath.com/IN-Grade8

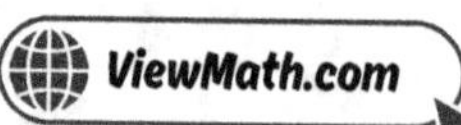

23. What is the volume of a sphere with radius 3 cm? Use $\pi \approx 3.14$.

   (A) $28.26\ cm^3$ (B) $113.04\ cm^3$

   (C) $36\pi\ cm^3$ (D) $84.78\ cm^3$

24. A square pyramid has volume $48\ in^3$ and height 4 in. What is the base side length?

Your Answer:

25. Can a scatter plot show a positive association that is nonlinear? Give an example.

Your Answer:

26. True or false: The line of best fit must pass through at least one data point.

   (A) True — it must go through the middle point.   (B) True — it must go through the first point.

   (C) False — it does not need to go through any   (D) False — it must stay above all data points.
data point.

27. A model $y = -2x + 50$ represents the number of pages left to read (y) after x hours. How long until the book is finished (all pages read)?

Your Answer:

28.

	Bike	Bus	Total
6th grade	18	22	40
7th grade	25	15	40
Total	43	37	80

What percentage of 6th graders ride the bus?

(A) 45%

(B) 55%

(C) 27.5%

(D) 62.5%

29. A spinner has sections: 30% red, 50% blue, 20% green. It is spun twice. Find P(blue, then green).

Your Answer

30. How many ways can 3 runners finish a race (1st, 2nd, 3rd) from a group of 8?

(A) 56

(B) 336

(C) 512

(D) 24

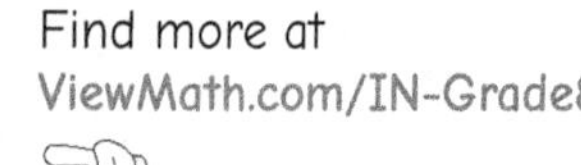
Find more at
ViewMath.com/IN-Grade8

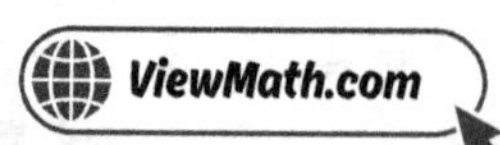

⭐ End of Practice Test 8 ⭐

Great job finishing the test!

 My Score

I got _____________ out of 30 questions right.

*Check your answers in the **Answer Key** at the back of the book.*

💡 *Review any questions you missed. That's how we learn!*

📊 Check Your Score Online!

Visit **ViewMath Academy** to enter your answers and see which topics you need to review. You can also explore lessons, take quizzes, track your scores, and save your progress!

viewmath.com/score/8.1.IN.23

Or go to viewmath.com/score and enter code: 8.1.IN.23

Practice Test 9

 30 Questions

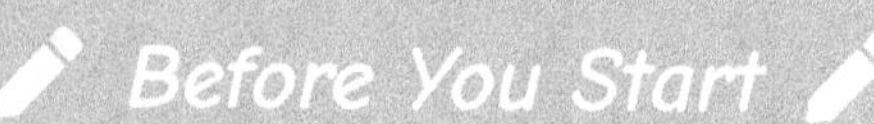

✏️ Before You Start ✏️

- ✔ **Read each question carefully** before choosing your answer.
- ✔ **Show your work** on scratch paper when you need to.
- ✔ **Skip hard questions** and come back to them later.
- ✔ **Check your answers** when you're done.
- ✔ **Take your time** — there's no rush!

 You've Got This! 

Do your best and show what you know!

1. Which of the following could NOT be the decimal expansion of a rational number?

(A) 2.500

(B) $0.\overline{9}$

(C) $1.41421356\ldots$ (non-repeating)

(D) $0.\overline{36}$

2. Between which two consecutive integers does $\sqrt{10}$ lie?

(A) 2 and 3

(B) 3 and 4

(C) 4 and 5

(D) 5 and 6

3. Between which two consecutive integers does $3\sqrt{2}$ lie?

(A) 3 and 4

(B) 4 and 5

(C) 5 and 6

(D) 6 and 7

4. Which expression equals $\frac{1}{9}$?

(A) 3^{-2}

(B) 3^{-3}

(C) 9^{-2}

(D) $(-3)^2$

5. Between which two consecutive whole numbers does $\sqrt{50}$ lie?

(A) 6 and 7

(B) 7 and 8

(C) 24 and 26

(D) 8 and 9

6. A factory makes 2.4×10^3 gadgets per hour. How many gadgets does it make in 5×10^2 hours?

(A) 1.2×10^5

(B) 1.2×10^6

(C) 12×10^5

(D) 7.4×10^3

Find more at
ViewMath.com/IN-Grade8

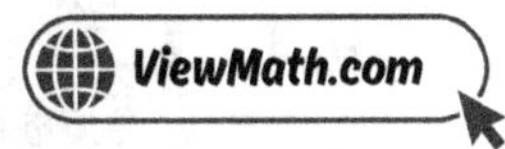

7. A proportional relationship passes through $(8, 12)$. What is y when $x = 20$?

(A) 24

(B) 28

(C) 30

(D) 32

8. The table shows a runner's distance over time.

Time (min)	0	5	10	15	20
Distance (m)	0	40	80	120	160

What is the runner's speed (slope of distance vs. time)?

Your Answer

9. Which line passes through the origin?

(A) $y = 2x + 1$

(B) $y = -3x$

(C) $y = x - 5$

(D) $y = 4x + 4$

10. A store sells T-shirts for \$12 and hats for \$8. Use the information in the table to find how many of each were sold.

	Value
Total items sold	15
Total revenue	\$148

Your Answer

Find more at
ViewMath.com/IN-Grade8

ViewMath.com

11. If $f(x) = -x + 8$, what is $f(-3)$?

(A) 5

(B) -5

(C) 11

(D) -11

12. Function A: $y = 5x + 3$. Function B: $y = 2x + 10$. Which function has a greater rate of change?

(A) Function A

(B) Function B

(C) They have the same rate of change.

(D) Cannot be determined.

13. Which of these is NOT a linear function?

(A) $y = 0$

(B) $y = 100x$

(C) $y = -x + 5$

(D) $y = \sqrt{x}$

14. A linear function passes through $(0, 5)$ and $(2, 11)$. What is its equation?

(A) $y = 2x + 5$

(B) $y = 3x + 5$

(C) $y = 5x + 3$

(D) $y = 6x + 5$

15. A decreasing graph does NOT necessarily mean:

(A) The output values are getting smaller.

(B) The graph goes downward from left to right.

(C) The output values are negative.

(D) The rate of change is negative.

16. After a translation, a triangle's longest side was originally 8 cm. What is the longest side of the image?

(A) 4 cm

(B) 8 cm

(C) 16 cm

(D) It depends on the direction of the translation.

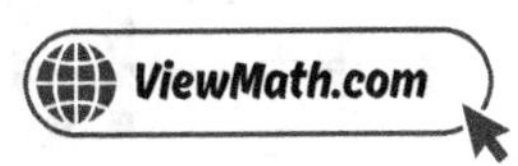

17. A pentagon is translated 4 units left and then rotated $90°$. Is the image congruent to the original?

(A) Yes

(B) No, because two transformations were used

(C) No, because it was rotated

(D) Only if it is a regular pentagon

18. Point $(-3, 5)$ is rotated $90°$ counterclockwise around the origin. What is its image?

(A) $(5, 3)$

(B) $(-5, -3)$

(C) $(3, -5)$

(D) $(-5, 3)$

19. A dilation with $k = 1$ produces:

(A) A figure twice as large

(B) A congruent figure

(C) A figure half as large

(D) No figure at all

20. Triangle ABC has $\angle A = 45°$ and $\angle B = 85°$. What type of triangle is it?

(A) Acute

(B) Right

(C) Obtuse

(D) Equilateral

21. Is the triangle with sides 9, 40, 41 a right triangle?

(A) No, because $9 + 40 \neq 41$.

(B) No, because $9^2 + 40^2 \neq 41^2$.

(C) Yes, because $9^2 + 40^2 = 41^2$.

(D) Yes, because $41 - 40 = 1$.

22. What is the distance between $(0, 0)$ and $(3, 4)$?

(A) 7

(B) 5

(C) $\sqrt{7}$

(D) 12

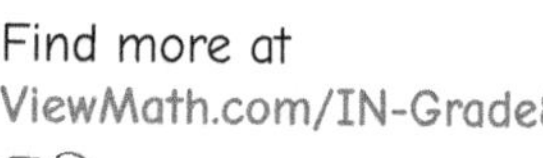

23. A cone has volume 80π cm^3 and radius 4 cm. What is the height?

(A) 5 cm

(B) 15 cm

(C) 10 cm

(D) 20 cm

24. A rectangular pyramid has a base of 8×5 m and a height of 12 m. What is its volume?

(A) 480 m^3

(B) 160 m^3

(C) 240 m^3

(D) 320 m^3

25. What type of association does this scatter plot show?

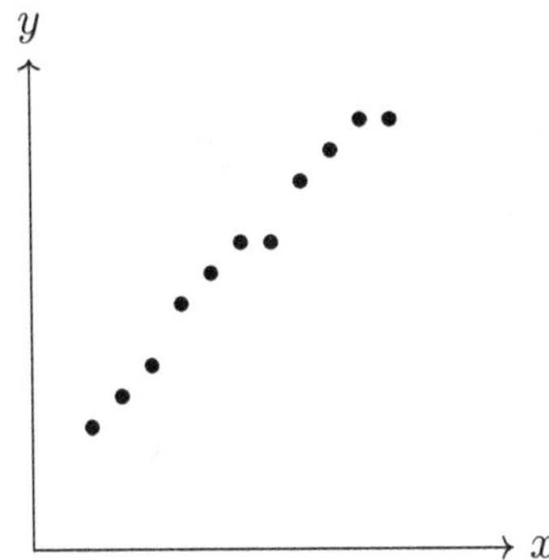

(A) Negative linear

(B) No association

(C) Positive linear

(D) Nonlinear

26. A trend line has equation $y = 1.5x + 2$. Using this, what is y when $x = 4$?

(A) 6

(B) 7

(C) 8

(D) 10

27. A scatter plot shows data and a trend line. The actual point at $x = 3$ is at $y = 8$, but the trend line passe
through $(3, 6)$. What is the residual?

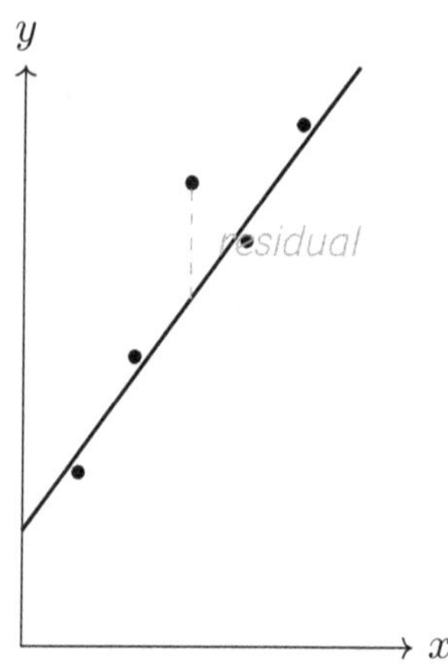

(A) -2

(B) 2

(C) 6

(D) 14

28. Using the same table, what percentage of students who did NOT study passed?

(A) 40%

(B) 60%

(C) 25%

(D) 50%

29. A coin is flipped 3 times. What is $P(\text{all heads})$?

(A) $\frac{3}{8}$

(B) $\frac{1}{8}$

(C) $\frac{1}{3}$

(D) $\frac{1}{6}$

30. A password must be 4 characters: first 2 are letters (A–Z, no repeat), last 2 are digits (0–9, no repeat). How many passwords?

Your Answer:

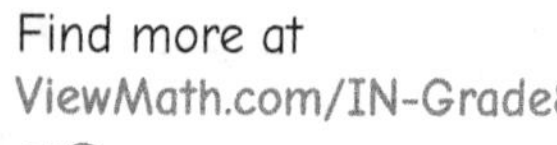

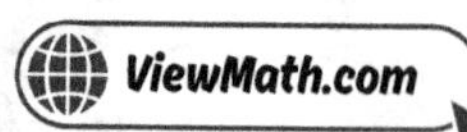

⭐ End of Practice Test 9 ⭐

Great job finishing the test!

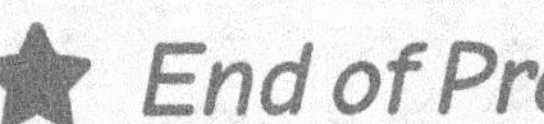 My Score

I got _____________ out of 30 questions right.

*Check your answers in the **Answer Key** at the back of the book.*

 *Review any questions you missed. That's how we learn!*

📊 Check Your Score Online!

Visit **ViewMath Academy** to enter your answers and see which topics you need to review. You can also explore lessons, take quizzes, track your scores, and save your progress!

viewmath.com/score/8.1.IN.24

Or go to viewmath.com/score and enter code: 8.1.IN.24

Practice Test 10

30 Questions

✏️ Before You Start ✏️

- ✓ **Read each question carefully** before choosing your answer.
- ✓ **Show your work** on scratch paper when you need to.
- ✓ **Skip hard questions** and come back to them later.
- ✓ **Check your answers** when you're done.
- ✓ **Take your time** — there's no rush!

⭐ You've Got This! ⭐

Do your best and show what you know!

1. Look at the Venn diagram below. In which region does $\sqrt{5}$ belong?

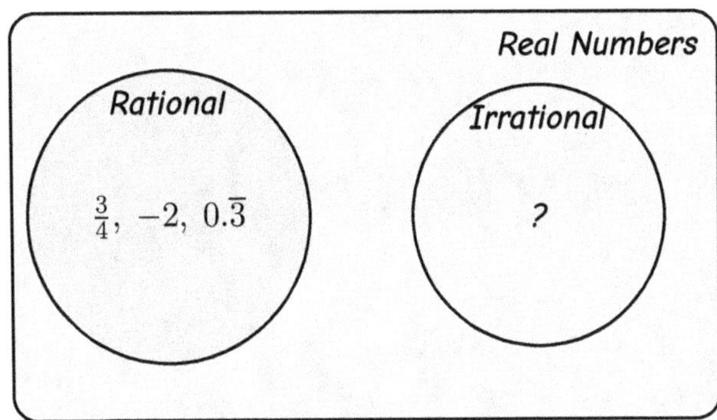

(A) Rational region, because $\sqrt{5} = 2.5$

(B) Irrational region, because 5 is not a perfect square

(C) Rational region, because $\sqrt{5}$ is a square root

(D) Neither region, because $\sqrt{5}$ is not a real number

2. Look at the number line below. Which point best represents $\sqrt{6}$?

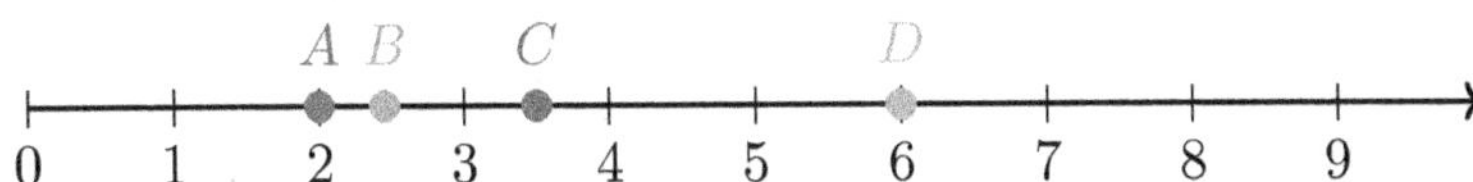

(A) Point A

(B) Point B

(C) Point C

(D) Point D

3. Which is the best estimate for π^2?

(A) 6.3

(B) 9.9

(C) 3.1

(D) 12.6

Find more at
ViewMath.com/IN-Grade8

4. Simplify $\left(\frac{3}{4}\right)^{-2}$.

   (A) $\frac{9}{16}$

   (C) $\frac{16}{9}$

   (B) $\frac{-9}{16}$

   (D) $\frac{-6}{8}$

5. Which statement about $\sqrt{2}$ is true?

   (A) $\sqrt{2}$ is a rational number

   (C) $\sqrt{2}$ is an irrational number

   (B) $\sqrt{2}$ is an integer

   (D) $\sqrt{2} = 1.5$

6. The diagram below shows the input and output of a machine that divides. Find the output value. Write your answer in scientific notation.

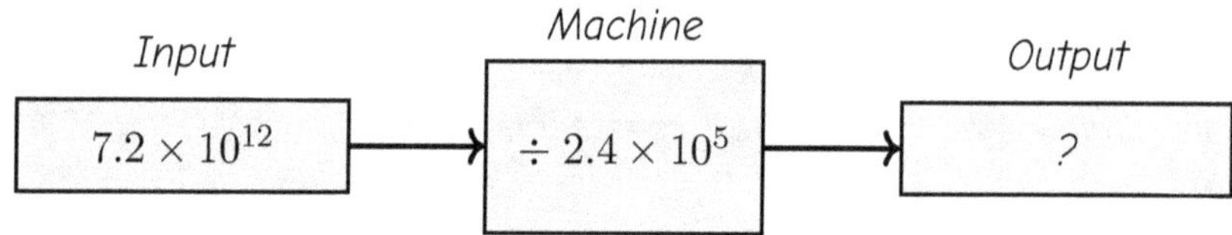

Your Answer:

7. A recipe uses 2 cups of flour for every 5 cookies. How many cups are needed for 30 cookies?

Your Answer:

8. Find the slope through $(3, -1)$ and $(7, 11)$.

Your Answer:

9. A line passes through the origin and $(3, 12)$. What is its equation?

(A) $y = 3x$

(B) $y = 12x$

(C) $y = 4x$

(D) $y = \frac{1}{4}x$

10. Pens cost \$2 and notebooks cost \$5. You buy 8 items for \$25. How many notebooks did you buy?

(A) 2

(B) 3

(C) 4

(D) 5

11. Use the graph of f below to find $f(1) + f(4)$.

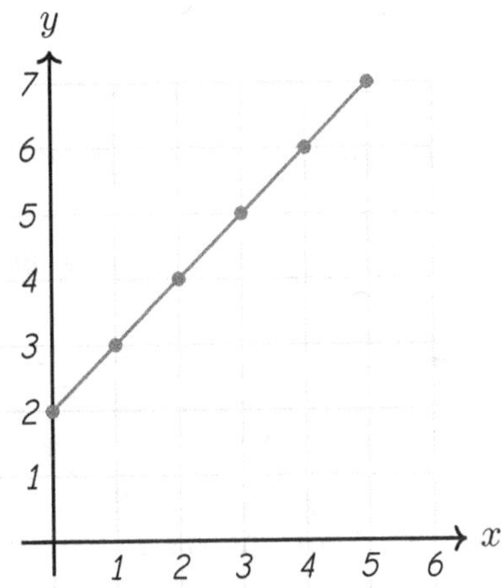

Your Answer:

12. Function A: $y = 2x + 100$. Function B: $y = 8x + 10$. At what value of x do the functions have the same output?

(A) $x = 10$

(B) $x = 15$

(C) $x = 18$

(D) $x = 20$

13. The table below shows a function. Calculate the first differences in y and determine whether the function is linear or nonlinear.

x	0	1	2	3	4
y	-3	1	5	9	13

Your Answer

14. Sam has \$120 and spends \$15 per day. Which function models the money remaining after x days?

(A) $y = 15x + 120$

(B) $y = -15x + 120$

(C) $y = 120x - 15$

(D) $y = -120x + 15$

15. Two distance-time graphs start at the same point. Graph A is steeper than Graph B. What can you say?

(A) Object A is farther from the start.

(B) Object A is traveling faster than Object B.

(C) Object B is traveling faster than Object A.

(D) Both objects travel at the same speed.

16. Point $M(5, -3)$ is reflected over the y-axis. What are the coordinates of M'?

Your Answer

17. A student says two circles with radius 5 cm are always congruent. Is this correct?

(A) No, circles cannot be congruent.

(B) No, they need the same center too.

(C) Yes, because they have the same radius.

(D) Yes, but only if they are in the same position.

Get Online

Find more at
ViewMath.com/IN-Grade8

ViewMath.com

18. *Point M is shown on the coordinate plane at $(-2, 5)$. Find the coordinates of M' after a 90° counterclockwise rotation around the origin.*

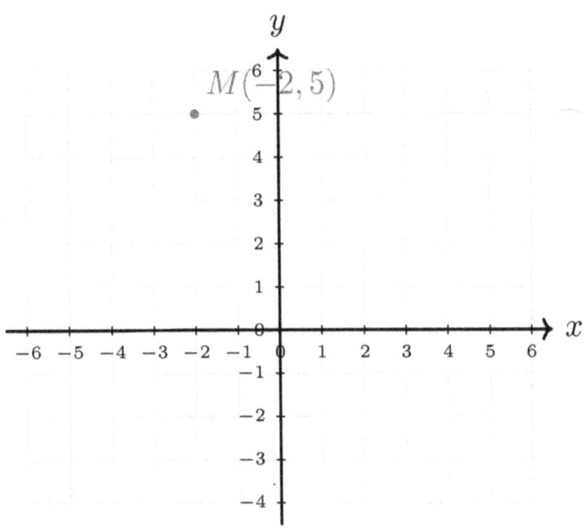

Your Answer:

19. *Two similar triangles have a scale factor of $\frac{1}{3}$ from the larger to the smaller. If the larger triangle's perimeter is 36 cm, what is the smaller triangle's perimeter?*

(A) 12 cm

(B) 108 cm

(C) 9 cm

(D) 18 cm

20. *Two parallel lines are cut by a transversal. If one co-interior (same-side interior) angle is 135°, what is the other?*

(A) 135°

(B) 45°

(C) 55°

(D) 225°

21. In a right triangle, the legs are 6 and 8. What is the hypotenuse?

   (A) 14   (B) 10

   (C) 48   (D) $\sqrt{48}$

22. Two points are at $(3, 7)$ and $(3, -5)$. What is the distance between them?

   (A) 2   (B) 12

   (C) -12 (D) $\sqrt{12}$

23. A cylindrical can has radius 3 in and height 8 in. How much soup can it hold? Use $\pi \approx 3.14$.

> Your Answer

24. A square pyramid and a rectangular prism share the same base (6×6 in) and height (10 in). What is the volume of the pyramid?

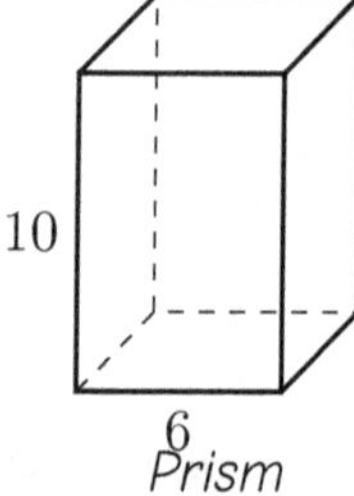
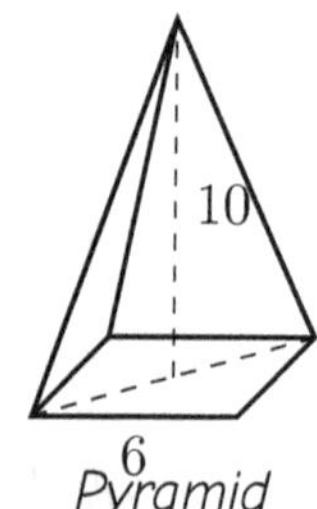

   (A) $360\ in^3$ (B) $180\ in^3$

   (C) $120\ in^3$ (D) $60\ in^3$

Find more at
ViewMath.com/IN-Grade8

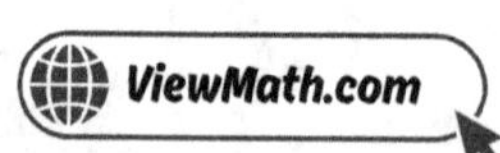

25. True or false: A scatter plot can show both clustering and outliers at the same time.

(A) True — they describe different features of the data.

(B) False — data has either clusters or outliers, not both.

(C) True — but only if there is no association.

(D) False — outliers are always in clusters.

26. Why should outliers generally be ignored when drawing a line of best fit?

Your Answer:

27. A linear model is $y = 3x + 10$, where x is hours studied and y is the test score. What does the y-intercept represent?

(A) The score after 3 hours of studying

(B) The score with 0 hours of studying

(C) The number of hours needed to pass

(D) The maximum possible score

28. The bar chart below shows survey results about favorite season by gender. Which statement is supported?

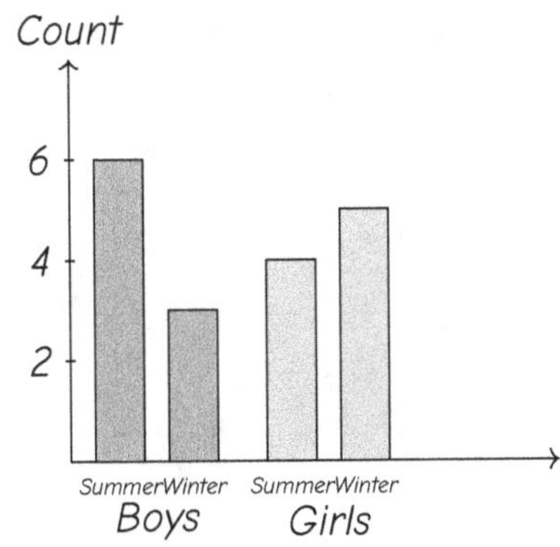

(A) Boys and girls prefer summer equally.

(B) Boys prefer summer more, girls prefer winter more.

(C) Girls prefer summer more than boys.

(D) There is no difference in preference.

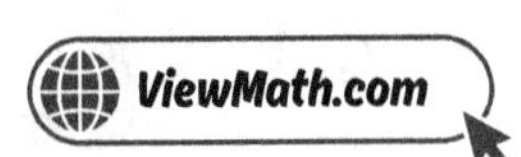

29. *A fair die is rolled twice. What is $P(\text{both rolls are } 6)$?*

(A) $\frac{1}{6}$ 　　　　　　　　　　　　(B) $\frac{1}{12}$

(C) $\frac{1}{36}$ 　　　　　　　　　　　(D) $\frac{2}{6}$

30. *You pick 2 books from a shelf of 5 to read (order doesn't matter). How many combinations?*

(A) 20 　　　　　　　　　　　　(B) 10

(C) 25 　　　　　　　　　　　　(D) 5

Great job finishing the test!

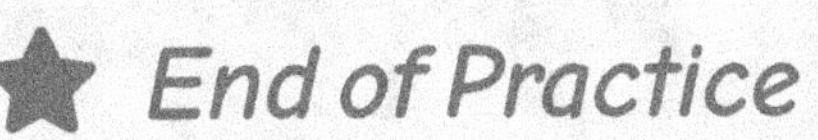 **My Score**

I got _____________ out of 30 questions right.

*Check your answers in the **Answer Key** at the back of the book.*

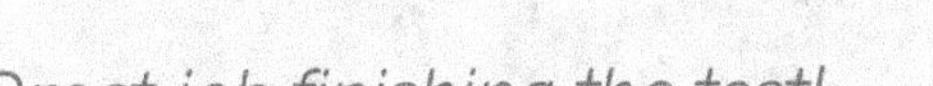 *Review any questions you missed. That's how we learn!*

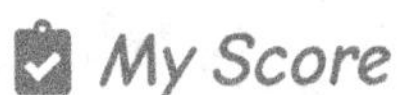 **Check Your Score Online!**

Visit **ViewMath Academy** to enter your answers and see which topics you need to review. You can also explore lessons, take quizzes, track your scores, and save your progress!

viewmath.com/score/8.1.IN.25

Or go to viewmath.com/score and enter code: 8.1.IN.25

Answer Key & Explanations

Answer Key

First try each test on your own, then check your work here.

✔ Practice Test 1 — Answer Key

1 False **2** A **3** B **4** D **5** B **6** 317 *times* **7** A **8** B

9 $y = 10x + 25$ **10** 80 **11** C

12 *Function A: rate* $= 3$, *initial* $= 20$. *Function B: rate* $= 5$, *initial* $= 5$. *At* $x = 10$, $B = 55$ *is greater than* $A = 50$.

13 Yes **14** B **15** Constant **16** C **17** Reflection over the y-axis **18** $(-5, -2)$

19 B **20** B **21** B **22** C **23** A **24** C **25** B **26** A **27** B

28 *Boys:* $\frac{20}{30} \approx 66.7\%$. *Girls:* $\frac{12}{20} = 60\%$. *The percentages are close, so association is weak or none.* **29** B

30 *Choosing 2 items to include is the same as choosing 3 items to exclude.* $\binom{n}{r} = \binom{n}{n-r}$.

💡 Time to Learn! 💡

*Review the explanations below, **especially for the questions you missed**.*

Understanding why each answer is correct builds stronger problem-solving skills.

***Tip:** Circle any questions you got wrong, then read their explanation carefully.*

📖 Practice Test 1 — Detailed Explanations

1 $\frac{22}{7} \approx 3.142857\ldots$ is a rational approximation of π, but $\pi = 3.14159265\ldots$ is irrational. They are close but not equal.

2 The point is between 5 and 6, closer to 5. $\sqrt{28} \approx 5.29$, which matches this position. $\sqrt{33} \approx 5.74$ would be closer to 6, and $\sqrt{40} \approx 6.32$ and $\sqrt{55} \approx 7.42$ would be beyond 6.

3 Area $= \sqrt{20} \times \sqrt{8} = \sqrt{160}$. $\sqrt{160} \approx 12.65$ cm^2. (Note: $\sqrt{160} = 4\sqrt{10} \approx 12.65$.) Choices A and C confuse addition with multiplication.

4 $7^5 \cdot 7^{-5} = 7^{5+(-5)} = 7^0 = 1$.

5 If $s = 8$, then $A = 8^2 = 64$. If $e = 4$, then $V = 4^3 = 64$. Since $A = V = 64$, the pair $s = 8, e = 4$ works.

6 $\frac{1.9 \times 10^{27}}{6 \times 10^{24}} = \frac{1.9}{6} \times 10^3 \approx 0.317 \times 10^3 = 317$.

7 Table: $k = \frac{6}{1} = 6$. Equation: $k = 5$. The table has the larger constant of proportionality.

8 The slope equals the cost per text, which is \$0.15.

9 Slope (rate) $= 10$ and y-intercept (initial fee) $= 25$: $y = 10x + 25$.

10 $a + s = 200$ and $8a + 5s = 1240$. From first: $s = 200 - a$. Substitute: $8a + 5(200 - a) = 1240$, so $3a + 1000 = 1240$, $3a = 240$, $a = 80$.

11 $g(x) = 0$ where the line crosses the x-axis. The graph crosses at $x = 4$.

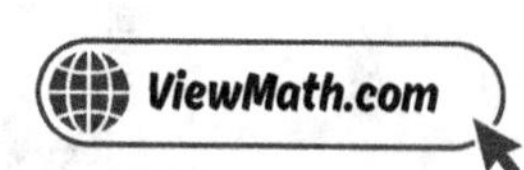

12 A: slope 3, $b = 20 \Rightarrow A(10) = 3(10) + 20 = 50$. B: slope 5, $b = 5 \Rightarrow B(10) = 5(10) + 5 = 55$. Function B is greater at $x = 10$.

13 $y = -12$ is a horizontal line, which can be written as $y = 0x + (-12)$ $(m = 0, b = -12)$. It is linear.

14 The rate is \$3 per mile $(m = 3)$ and the flat fee is \$4 $(b = 4)$. So $y = 3x + 4$.

15 A straight line has a constant slope, meaning the rate of change is the same everywhere along the line.

16 A translation moves the figure to a new position. Side lengths, angle measures, and parallelism are all preserved.

17 Each x-coordinate is negated while y stays the same: $(1,1) \to (-1,1)$, etc. This is a reflection over the y-axis.

18 $180°$ rotation: $(x, y) \to (-x, -y)$. So $(5, 2) \to (-5, -2)$.

19 Divide the image coordinates by the original: $\frac{12}{4} = 3$ and $\frac{-6}{-2} = 3$. The scale factor is 3.

20 Alternate interior angles are equal: $4x = 80$, so $x = 20$.

21 $c^2 = 5^2 + 5^2 = 25 + 25 = 50$, so $c = \sqrt{50}$.

22 $d = \sqrt{(6 - 2)^2 + (-2 - (-5))^2} = \sqrt{16 + 9} = \sqrt{25} = 5$.

23 A cone is exactly $\frac{1}{3}$ the volume of a cylinder with the same base and height: $\frac{450\pi}{3} = 150\pi$ cm^3.

24 Prism volume $= 96$ cm^3. Pyramid volume $= \frac{1}{3}(16)(6) = 32$ cm^3. $96/32 = 3$.

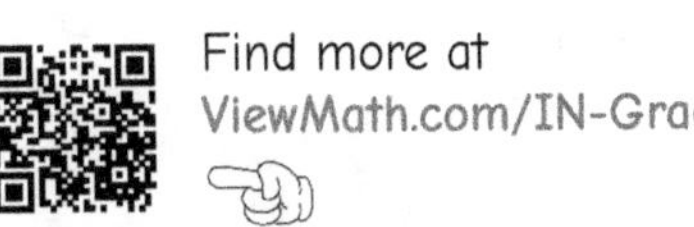

Find more at
ViewMath.com/IN-Grade8

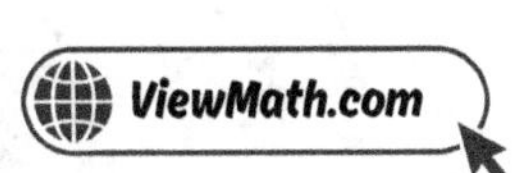

25 One variable goes up while the other goes down, forming a negative association.

26 Slope $= \frac{7-3}{5-1} = 1$. y-intercept: $3 = 1(1) + b$, $b = 2$. Equation: $y = x + 2$.

27 The linear pattern observed in data may not hold outside the data range. Real-world relationships can change.

28 Compare the conditional relative frequencies. Since 66.7% and 60% are similar, the preference doesn't change much by gender.

29 $P(\text{not event}) = 1 - P(\text{event}) = 1 - \frac{3}{4} = \frac{1}{4}$.

30 $\binom{5}{2} = 10$ and $\binom{5}{3} = 10$. Each way of selecting 2 items corresponds to leaving out 3 items.

☑ Practice Test 2 — Answer Key

1 C
2 $\sqrt{5} \approx 2.2$, placed at the 2nd tick after 2.0
3 ≈ 0.02
4 D
5 B
6 D

7 C
8 0.1 miles per minute
9 A
10 5
11 B
12 7
13 Nonlinear
14 4

15 C
16 C
17 B
18 A
19 108 cm²
20 B
21 B
22 9
23 C

24 100 cm³
25 C
26 Approximate line through $(1, 2)$ and $(7, 9)$: slope $\approx \frac{7}{6} \approx 1.17$.
27 B

28 D
29 C
30 B

Find more at
ViewMath.com/IN-Grade8

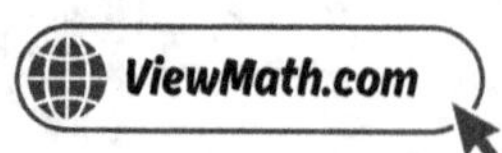

💡 Time to Learn! 💡

Review the explanations below, *especially for the questions you missed.*

Understanding why each answer is correct builds stronger problem-solving skills.

Tip: Circle any questions you got wrong, then read their explanation carefully.

📖 Practice Test 2 — Detailed Explanations

1. $P = \sqrt{2}$ is irrational (since 2 is not a perfect square) and $Q = 3$ is an integer, which is rational.

2. $2.2^2 = 4.84$ and $2.3^2 = 5.29$. Since 5 is closer to 4.84, $\sqrt{5} \approx 2.24$, which is near the 2nd to 3rd tick mark after 2.0.

3. $\pi \approx 3.1416$ and $\sqrt{10} \approx 3.1623$. The distance is $3.1623 - 3.1416 = 0.0207 \approx 0.02$.

4. A negative exponent means reciprocal: $4^{-3} = \frac{1}{4^3} = \frac{1}{64}$.

5. $x = \pm\sqrt{\frac{9}{16}} = \pm\frac{3}{4}$, since $\left(\frac{3}{4}\right)^2 = \frac{9}{16}$.

6. $9.1 - 8.6 = 0.5$, so 0.5×10^5. In proper notation: 5×10^4. Both representations are equivalent.

7. $k = \frac{y}{x} = \frac{20}{5} = 4$.

8. $m = \frac{5-2}{50-20} = \frac{3}{30} = 0.1$ miles per minute.

9. $m = \frac{11-2}{4-1} = 3$. Using $(1, 2)$: $2 = 3(1) + b$, so $b = -1$. Equation: $y = 3x - 1$.

Find more at
ViewMath.com/IN-Grade8

ViewMath.com

10 $a + b = 12$ and $5a + 10b = 85$. From first: $a = 12 - b$. Substitute: $5(12 - b) + 10b = 85$, so $60 + 5b = 85$, $5b = 25$, $b = 5$.

11 $f(2) = 10 - 3(2) = 10 - 6 = 4$ pieces remaining.

12   The initial value is the output when $x = 0$, which is 7.

13   Check the y-differences: $2 - 1 = 1$, $4 - 2 = 2$, $5 - 4 = 1$, $7 - 5 = 2$. The differences are not all the same $(1, 2, 1, 2)$, so the rate of change is not constant. The function is nonlinear.

14 $\frac{25-9}{6-2} = \frac{16}{4} = 4$.

15   A straight line means linear. Going upward means increasing.

16   Reflecting over the y-axis flips the sign of the x-coordinate while keeping y the same: $(7, -2) \rightarrow (-7, -2)$.

17   Equal angles make triangles similar, but not necessarily congruent — their sides could be different lengths.

18   First translate: $(1 + 3, 4 + (-2)) = (4, 2)$. Then reflect over the y-axis: $(4, 2) \rightarrow (-4, 2)$.

19   Area scales by k^2: $12 \times 3^2 = 12 \times 9 = 108$ cm^2.

20   Alternate interior angles formed by parallel lines and a transversal are equal: $72°$.

21 $c^2 = 8^2 + 15^2 = 64 + 225 = 289$, so $c = \sqrt{289} = 17$.

22   Same x-coordinates, so distance $= |7 - (-2)| = 9$.

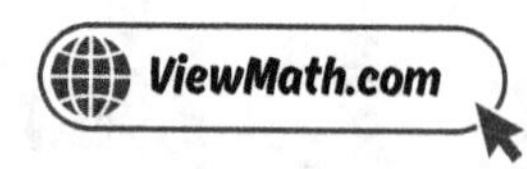

23. $V = \pi r^2 h = \pi(49)(3) = 147\pi \ m^3$.

24. Base area $= \frac{1}{2}(5)(12) = 30 \ cm^2$. $V = \frac{1}{3}(30)(10) = 100 \ cm^3$.

25. Point P at $(3, 7)$ is far above the general upward trend around $y \approx 2.5$ when $x = 3$. It is the outlier.

26. $m = \frac{9-2}{7-1} = \frac{7}{6} \approx 1.17$. The line rises about 1.2 units for each 1-unit increase in x.

27. In $y = mx + b$, the slope m represents the rate of change: how much y changes for each 1-unit increase in x.

28. Joint frequencies fill the interior cells: $3 \times 4 = 12$.

29. 2 coin outcomes $\times$ 6 die outcomes $= 12$ total outcomes.

30. $3! = 6$. The tree shows all 6 arrangements: ABC, ACB, BAC, BCA, CAB, CBA.

✔ Practice Test 3 — Answer Key

 C
 C
 C
 B
 C
 A
 A
 A
 C
 C

 B
 B
 B
 $y = 4x + 4$
 B
 B
 A
 A
 C

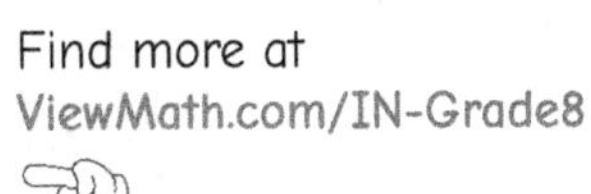 B
 B
22. 10
23. C
24. A
25. C
26. Approximately 2 to 2.25
27. B

28. B
29. B
30. B

Get Online

Find more at
ViewMath.com/IN-Grade8

ViewMath.com

💡 Time to Learn! 💡

*Review the explanations below, **especially for the questions you missed**.*

Understanding why each answer is correct builds stronger problem-solving skills.

Tip: *Circle any questions you got wrong, then read their explanation carefully.*

📖 Practice Test 3 — Detailed Explanations

1 $0.\overline{81}$ is a repeating decimal, so it can be written as a fraction: $0.\overline{81} = \frac{81}{99} = \frac{9}{11}$. The others are irrational.

2 $5.4^2 = 29.16$ and $5.5^2 = 30.25$. Since 30 is very close to 30.25, $\sqrt{30} \approx 5.5$.

3 $4\sqrt{2} \approx 4 \times 1.414 = 5.656$ and $\sqrt{30} \approx 5.477$. Since $5.656 > 5.477$, $4\sqrt{2}$ is larger.

4 $\frac{8^6}{8^6} = 8^{6-6} = 8^0 = 1$. The expression equals 8^0.

5   Side $= \sqrt{196} = 14$ cm, since $14 \times 14 = 196$.

6   Divide the coefficients: $\frac{8}{2} = 4$. Subtract the exponents: $10^{7-3} = 10^4$. Answer: 4×10^4.

7   Machine A: 12 bottles/hour. Machine B: $\frac{50}{5} = 10$ bottles/hour. Machine A is faster.

8 $m = \frac{1-7}{5-2} = \frac{-6}{3} = -2$.

9   Steepness is $|m|$: $|1| = 1$, $|3| = 3$, $|-5| = 5$, $\left|\frac{1}{2}\right| = 0.5$. The steepest has $|m| = 5$.

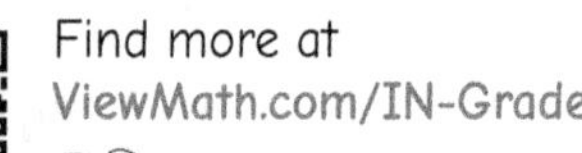
Find more at
ViewMath.com/IN-Grade8

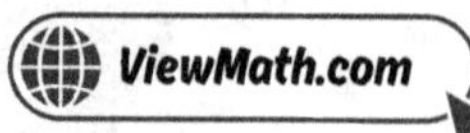

10 $60t + 80t = 420$, so $140t = 420$, $t = 3$ hours.

11 $f(0) = -1$ and $f(2) = 7$. So $f(0) + f(2) = -1 + 7 = 6$.

12 $|-4| = 4 > |-1| = 1$, so Function B decreases faster.

13 $7 - 3 = 4$, $11 - 7 = 4$, $15 - 11 = 4$. The constant change of 4 proves this is linear.

14 Slope: $\frac{24-4}{5-0} = \frac{20}{5} = 4$. y-intercept: 4. Equation: $y = 4x + 4$.

15 A downward straight line means the quantity decreases at a constant rate — water is leaking steadily.

16 The 90° CCW rule is $(x, y) \to (-y, x)$. So $(-1, 4) \to (-4, -1)$.

17 Each vertex of Figure A is shifted 6 units right to get Figure B: $(1, 1) \to (7, 1)$, $(4, 1) \to (10, 1)$, etc. Since a translation preserves size and shape, the quadrilaterals are congruent.

18 90° CCW: $(x, y) \to (-y, x)$. So $(0, -5) \to (5, 0)$.

19 Multiply the side by the scale factor. $3 \times 4 = 12$ cm.

20 $180 - 54 - 54 = 72°$.

21 The diagonal of a square with side s is $s\sqrt{2}$. So the distance is $90\sqrt{2} \approx 127.3$ ft.

22 Horizontal: $4 - (-4) = 8$. Vertical: $4 - (-2) = 6$. $d = \sqrt{8^2 + 6^2} = \sqrt{64 + 36} = \sqrt{100} = 10$.

23 $V = \frac{4}{3}\pi(3r)^3 = \frac{4}{3}\pi(27r^3) = 27 \times \frac{4}{3}\pi r^3$. The volume is 27 times as large.

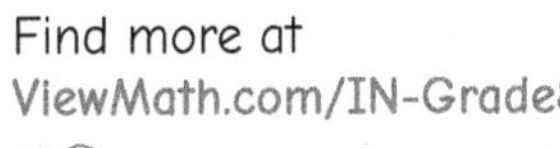
Find more at
ViewMath.com/IN-Grade8

24 $V = \frac{1}{3}(49)(12) = \frac{588}{3} = 196 \ cm^3$.

25 A strong positive trend corresponds to an r-value close to 1. $r = 0.9$ indicates a strong positive association.

26 Using endpoints: $m \approx \frac{11-2}{5-1} = \frac{9}{4} = 2.25$. A slope around 2 is a reasonable estimate.

27 Residual = actual − predicted. A positive residual means the actual value is above the predicted value.

28 The 20 is the count for a specific combination of two variables (7th grade AND soccer), making it a joint frequency.

29 Independent events have no effect on each other. The occurrence of one does not change the probability of the other.

30 $4 \times 3 \times 2 = 24$. Order matters and no repeats.

✅ Practice Test 4 — Answer Key

 D  ≈ 8.2  B  1.2  B  A  D  0  B

 C  C  $A = 21, B = 21$  C  $y = 3x$  C  B  C

 B  A 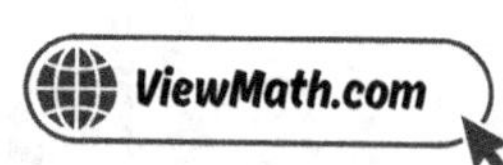 A **21** B **22** C **23** B **24** 70 cm^3 **25** D **26** B

27 B **28** B **29** $P(both\ blue) = \frac{6}{10} \times \frac{5}{9} = \frac{30}{90} = \frac{1}{3}$. **30** B

Find more at
ViewMath.com/IN-Grade8

💡 Time to Learn! 💡

Review the explanations below, **especially for the questions you missed**.

Understanding why each answer is correct builds stronger problem-solving skills.

Tip: Circle any questions you got wrong, then read their explanation carefully.

📖 Practice Test 4 — Detailed Explanations

1 $\sqrt{7}$ is irrational because 7 is not a perfect square. The other choices are all rational: $\frac{5}{8} = 0.625$, $0.\overline{6} = \frac{2}{3}$, and $\sqrt{9} = 3$.

2 $8.2^2 = 67.24$ and $8.3^2 = 68.89$. Since 68 is between these and closer to 68.89, $\sqrt{68} \approx 8.2$.

3 $2\sqrt{5} = \sqrt{4} \cdot \sqrt{5} = \sqrt{4 \times 5} = \sqrt{20}$. Alternatively, $2\sqrt{5} \approx 4.47$ while $\sqrt{10} \approx 3.16$. They are not equal.

4 $5^0 = 1$ and $5^{-1} = \frac{1}{5} = 0.2$. So $1 + 0.2 = 1.2$.

5 $7 \times 7 = 49$, so $\sqrt{49} = 7$.

6 Same exponent, so add the coefficients: $5.2 + 3.8 = 9.0$. Answer: 9×10^6.

7 If the relationship passes through $(1, 5)$ and $(3, 15)$, then $k = 5$. But $5 \times 2 = 10 \neq 13$, so $(2, 13)$ does not fit.

8 $m = \frac{4-4}{8-0} = \frac{0}{8} = 0$. The line is horizontal.

9 $(0, -4)$ gives $b = -4$. With $m = \frac{2}{3}$: $y = \frac{2}{3}x - 4$.

10 $x + y = 25$ and $x - y = 7$. Add: $2x = 32$, $x = 16$. The larger number is 16.

11 $h(5) = 2(5)^2 = 2(25) = 50$.

12 $A(3) = 2(3) + 15 = 21$. $B(3) = 5(3) + 6 = 21$. They are equal at $x = 3$.

13 $2 - 1 = 1$, $5 - 2 = 3$, $10 - 5 = 5$, $17 - 10 = 7$. The differences change, so the rate of change is not constant — nonlinear.

14 Slope: $\frac{12-3}{4-1} = \frac{9}{3} = 3$. Using $(1, 3)$: $3 = 3(1) + b \Rightarrow b = 0$. Equation: $y = 3x$.

15 Filling: water level rises (increasing). Soaking: level stays the same (constant). Draining: level drops (decreasing).

16 Each vertex has its x-coordinate negated while y stays the same: $P(1, 1) \to P'(-1, 1)$, $Q(4, 1) \to Q'(-4, 1)$, $R(2, 4) \to R'(-2, 4)$. This is a reflection over the y-axis.

17 Both rectangles have dimensions 6×10. A rotation maps one onto the other, so they are congruent.

18 Add the translation: $(-6 + 4,\ 2 + (-5)) = (-2, -3)$.

19 Congruent figures are a special case of similar figures with $k = 1$. All congruent figures are similar, but not all similar figures are congruent.

20 An exterior angle and its adjacent interior angle are supplementary: $180 - 140 = 40°$.

21 $7^2 + 24^2 = 49 + 576 = 625 = 25^2$. Since $a^2 + b^2 = c^2$, it is a right triangle.

22 $d = \sqrt{(2 - (-1))^2 + (6 - 2)^2} = \sqrt{9 + 16} = \sqrt{25} = 5$.

Find more at
ViewMath.com/IN-Grade8

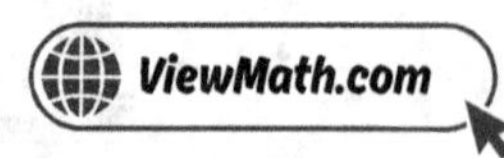

23. $V = \frac{1}{3}\pi(4^2)(9) = \frac{1}{3}\pi(144) = 48\pi \ cm^3$.

24. $V = \frac{1}{3}(35)(6) = \frac{210}{3} = 70 \ cm^3$.

25. If there is no clear trend or pattern, there is no association between the variables.

26. $m = \frac{19-7}{9-3} = \frac{12}{6} = 2$.

27. Predicted: $y = 2.5(4) + 1 = 11$. Residual $=$ actual $-$ predicted $= 12 - 11 = 1$.

28. Total who passed $= 40$. Total students $= 60$. Fraction $= \frac{40}{60} = \frac{2}{3}$.

29. First blue: $\frac{6}{10}$. Second blue (without replacement): $\frac{5}{9}$. $P = \frac{6}{10} \times \frac{5}{9} = \frac{30}{90} = \frac{1}{3}$.

30. $\binom{10}{2} = \frac{10\times 9}{2} = 45$. Order doesn't matter.

✅ Practice Test 5 — Answer Key

1. Rational   2. B   3. A   4. B   5. C   6. 4×10^{-6} 7. A   8. A   9. C

10. Length $= 16$ m, width $= 8$ m   11. 19   12. A   13. Nonlinear; each output is x^3.   14. B

15. A   16. $(0, -4)$ 17. C   18. C   19. C   20. B   21. 26   22. A   23. B

24. B   25. C   26. B

27. At $x = 5$: $y = 30$. At $x = 30$: $y = -45$. The prediction at $x = 5$ is more reliable because it is within the data range; $x = 3$

28. B   29. Outcomes: HH, HT, TH, TT. P(at least one H) $= \frac{3}{4}$.   30. B

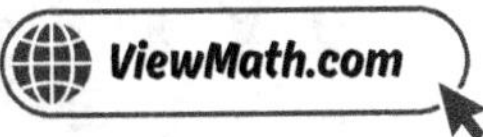

> ### 💡 Time to Learn! 💡
>
> Review the explanations below, **especially for the questions you missed.**
>
> Understanding why each answer is correct builds stronger problem-solving skills.
>
> **Tip:** Circle any questions you got wrong, then read their explanation carefully.

📖 Practice Test 5 — Detailed Explanations

1. $\sqrt{144} = 12$ exactly, because $12^2 = 144$. Since 12 is an integer, $\sqrt{144}$ is rational.

2. $7^2 = 49$ and $8^2 = 64$. Since $49 < 50 < 64$, we have $7 < \sqrt{50} < 8$.

3. Perimeter $= 4 \times \sqrt{50} = 4\sqrt{50} \approx 4 \times 7.07 = 28.28 \approx 28.3$ cm. Note that $\sqrt{200} \approx 14.14$, which is only $2\sqrt{50}$ — that would be two sides, not four.

4. The star is at $\frac{1}{8}$. Since $\frac{1}{8} = \frac{1}{2^3} = 2^{-3}$, the answer is 2^{-3}.

5. $81 = 9^2$, so 81 is a perfect square.

6. $8 \times 5 = 40$ and $10^{-4} \times 10^{-3} = 10^{-7}$. Then $40 \times 10^{-7} = 4 \times 10^{-6}$.

7. Store A: $k = 2$ per pound. Store B: $k = \frac{7.50}{3} = 2.50$ per pound. Store A is cheaper.

8. $\frac{10-40}{15} = \frac{-30}{15} = -2$ gal/min. The negative sign shows the water level is decreasing.

9. $m = \frac{8-(-2)}{5-0} = \frac{10}{5} = 2$. The y-intercept is -2: $y = 2x - 2$.

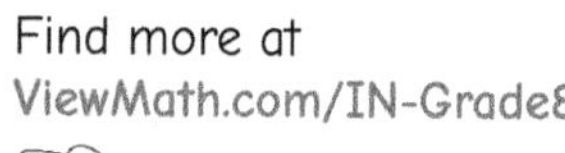
Find more at
ViewMath.com/IN-Grade8

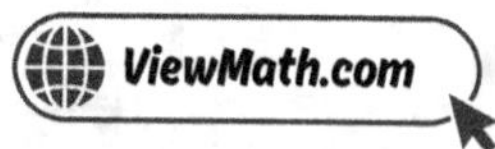

10 $2l + 2w = 48$ and $l = 2w$. Substitute: $2(2w) + 2w = 48$, $6w = 48$, $w = 8$. Then $l = 16$.

11 $f(5) = 2(5) + 9 = 10 + 9 = 19$.

12   Function P's rate of change: $\frac{7-4}{1-0} = 3$. Function Q's slope is 2. Since $3 > 2$, Function P grows faster.

13 $1^3 = 1$, $2^3 = 8$, $3^3 = 27$, $4^3 = 64$. The differences $(7, 19, 37)$ are not constant, confirming nonlinear.

14 $y = 25(8) + 60 = 200 + 60 = 260$.

15   The ball goes up, reaches a peak, then comes down. Gravity causes the rate of change to vary, so the path is curved (nonlinear).

16   A $180°$ rotation uses $(x, y) \to (-x, -y)$. So $(0, 4) \to (0, -4)$.

17   In $\triangle ABC$: $\angle C = 180 - 55 - 80 = 45°$. Since $\angle C$ corresponds to $\angle F$ in congruent triangles, $\angle F = 45°$.

18   Reflecting over the y-axis negates x and keeps y: $(-4, 2) \to (4, 2)$.

19   All squares have four $90°$ angles. Any two squares can be mapped by a dilation since the ratio of their sides is constant. So all squares are similar.

20   Corresponding angles are in the same position (e.g., both upper-left) at each intersection of the transversal with the parallel lines.

21 $c^2 = 10^2 + 24^2 = 100 + 576 = 676$, so $c = \sqrt{676} = 26$.

22 $d = \sqrt{(4 - (-2))^2 + (9 - 1)^2} = \sqrt{6^2 + 8^2} = \sqrt{36 + 64} = \sqrt{100} = 10$.

23 $V = \frac{1}{3}(3.14)(64)(24) = \frac{1}{3}(4{,}823.04) \approx 1{,}607.7\ in^3.$

24 $V = \frac{1}{3}Bh = \frac{1}{3}(36)(10) = 120\ cm^3.$

25   A linear association means the data points roughly follow a straight-line trend.

26   A good fit means the data points are close to the line, with small distances from each point to the line.

27 $y = -3(5)+45 = 30$ (interpolation). $y = -3(30)+45 = -45$ (extrapolation — unreliable and gives a negative value that may not make sense).

28 $0.15 \times 200 = 30.$

29   4 outcomes total. Only TT has no heads. $P(\text{at least one } H) = 1 - P(TT) = 1 - \frac{1}{4} = \frac{3}{4}.$

30 $\binom{6}{3} = \frac{6!}{3!\cdot 3!} = \frac{720}{6\cdot 6} = 20.$

✅ Practice Test 6 — Answer Key

1 Rational	2 A	3 D	4 C	5 A	6 B	7 C	8 B	9 B
10 C	11 A	12 A	13 Linear	14 B	15 B	16 C	17 5 cm	18 $(-3,8)$
19 B	20 40	21 B	22 $\sqrt{13}$	23 B	24 B	25 C	26 C	27 C	28 B
29 B	30 B							

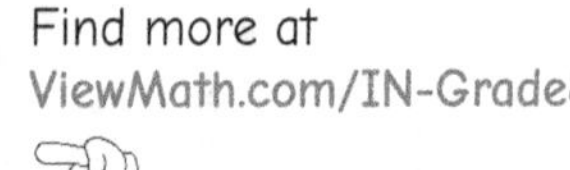
Find more at
ViewMath.com/IN-Grade8

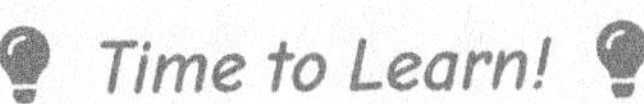

💡 Time to Learn! 💡

Review the explanations below, *especially for the questions you missed*.

Understanding why each answer is correct builds stronger problem-solving skills.

Tip: Circle any questions you got wrong, then read their explanation carefully.

📖 Practice Test 6 — Detailed Explanations

1. $\frac{5}{6}$ is a fraction of two integers with a nonzero denominator, so it is rational. Its decimal is $0.8\overline{3}$, which repeats.

2. $6^2 = 36$ and $7^2 = 49$. Since $36 < 45 < 49$, we have $6 < \sqrt{45} < 7$. The student is correct.

3. $\pi \approx 3.1416$, $\sqrt{10} \approx 3.1623$, and 3.2 is exact. From least to greatest: $3.1416 < 3.1623 < 3.2$.

4. Any nonzero number raised to the zero power equals 1. So $(-2)^0 = 1$.

5. Edge $= \sqrt[3]{343} = 7$ in., since $7^3 = 343$.

6. Same exponent, so subtract the coefficients: $7.5 - 2.5 = 5$. Answer: 5×10^8.

7. $k = \frac{18}{6} = 3$, so the equation is $y = 3x$.

8. $\frac{270-120}{5-2} = \frac{150}{3} = 50$ mph.

9. Substitute $m = -2$ and $b = 7$ into $y = mx + b$: $y = -2x + 7$.

Find more at
ViewMath.com/IN-Grade8

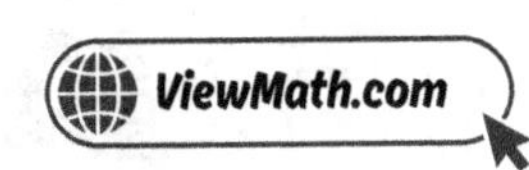

10 $n + d = 15$ and $5n + 10d = 110$. From first: $n = 15 - d$. Substitute: $5(15 - d) + 10d = 110$, so $75 + 5d = 110$, $5d = 35$, $d = 7$.

11 $f(10) = 7 - 10 = -3$.

12 Function A is steeper — it rises faster. Its slope is 2 compared to Function B's slope of 1.

13 A constant increase of 6 means the rate of change is constant, which is the defining property of a linear function.

14 Starting height $b = 12$. It loses 2 inches per hour, so $m = -2$. The function is $y = -2x + 12$.

15 A steeper upward slope means faster increase. The morning section is steeper than the afternoon section, so the temperature rose faster in the morning.

16 Reflections are rigid transformations that preserve side lengths, angles, area, and perimeter.

17 $P = 2l + 2w$. So $28 = 2(9) + 2w$, giving $28 = 18 + 2w$, $2w = 10$, $w = 5$ cm.

18 Subtract the translation: $a = 2 - 5 = -3$, $b = 7 - (-1) = 8$. So $(a, b) = (-3, 8)$.

19 Similar figures have the same shape, which means all corresponding angles are equal. Side lengths, perimeter, and area may differ.

20 $90 + x + (x + 10) = 180$. Simplify: $2x + 100 = 180$, $2x = 80$, $x = 40$.

21 $c^2 = 1^2 + 1^2 = 2$, so $c = \sqrt{2}$.

22 $d = \sqrt{2^2 + 3^2} = \sqrt{4 + 9} = \sqrt{13}$.

Find more at
ViewMath.com/IN-Grade8

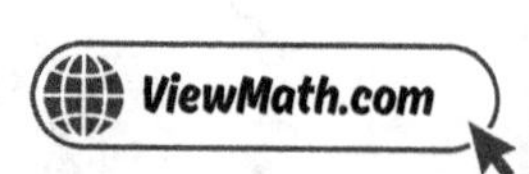

23. $V = \frac{1}{3}\pi r^2 h = \frac{1}{3}\pi(9)(12) = \frac{108\pi}{3} = 36\pi \; m^3.$

24. $V = \frac{1}{3}Bh = \frac{1}{3}(30)(9) = 90 \; cm^3.$

25. The independent (explanatory) variable goes on the x-axis; the dependent (response) variable goes on the y-axis.

26. The y-intercept is the y-value when $x = 0$. The line passes through $(0, 3)$, so $b = 3$.

27. A slope of 0 means no change in y as x changes; the line is horizontal.

28. Relative frequency is the ratio of a frequency to the total, often written as a fraction, decimal, or percentage.

29. $Total = 3 + 5 + 2 = 10.$ $P(blue) = \frac{5}{10} = \frac{1}{2}.$

30. The Fundamental Counting Principle says you multiply the number of outcomes for each stage: $m \times n.$

✔ Practice Test 7 — Answer Key

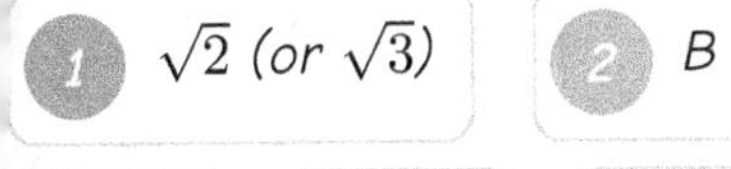 $\sqrt{2}$ (or $\sqrt{3}$)  B 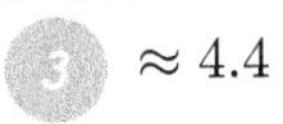 ≈ 4.4  C  6 cm  A 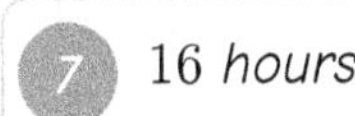 16 hours  B

 B  C  5  C  C  B  Decreasing  D  C

 B  B  25  B  D  $180\pi \; cm^3$  B

 No. Association (correlation) does not prove causation.  B

27. Predicted height at day 10: $y = 0.4(10) + 2 = 6$ cm. Slope: the plant grows 0.4 cm per day. y-intercept: the plant starts

Find more at
ViewMath.com/IN-Grade8

 B    B    C

💡 Time to Learn! 💡

Review the explanations below, **especially for the questions you missed.**

Understanding why each answer is correct builds stronger problem-solving skills.

Tip: Circle any questions you got wrong, then read their explanation carefully.

📖 Practice Test 7 — Detailed Explanations

1 $\sqrt{2} \approx 1.414$ is between 1 and 2 and is irrational because 2 is not a perfect square. $\sqrt{3} \approx 1.732$ also works.

2 $\sqrt{5} \approx 2.236$ and $\frac{5}{2} = 2.5$. From least to greatest: $2 < 2.236 < 2.5$.

3 $\sqrt{3} \approx 1.732$ and $\sqrt{7} \approx 2.646$. Adding: $1.732 + 2.646 = 4.378 \approx 4.4$.

4 Numerator: $9^3 \cdot 9^2 = 9^5$. Then $\frac{9^5}{9^4} = 9^{5-4} = 9^1 = 9$.

5 Edge $= \sqrt[3]{216} = 6$ cm, since $6 \times 6 \times 6 = 216$.

6 Multiply the coefficients: $3 \times 2 = 6$. Add the exponents: $10^{4+5} = 10^9$. So $(3 \times 10^4)(2 \times 10^5) = 6 \times 10^9$.

7 Rate $= \frac{6}{3} = 2$ hours per room. For 8 rooms: $2 \times 8 = 16$ hours.

8 $m = \frac{5-3}{5-1} = \frac{2}{4} = \frac{1}{2}$.

Find more at
ViewMath.com/IN-Grade8

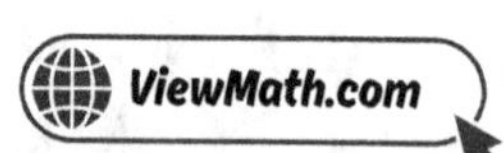

9. In $y = mx + b$, $m = 5$ is the slope and $b = -3$ is the y-intercept.

10. $2l + 2w = 34$ and $l = w + 5$. Substitute: $2(w + 5) + 2w = 34$, so $4w + 10 = 34$, $4w = 24$, $w = 6$.

11. $f(6) = \frac{6+4}{2} = \frac{10}{2} = 5$.

12. $\frac{9-1}{2-0} = \frac{8}{2} = 4$. The rate of change is 4.

13. $A = s^2$ has the variable raised to the 2nd power. For a function to be linear, the variable must have an exponent of 1.

14. Slope: $\frac{20-8}{3-0} = \frac{12}{3} = 4$. y-intercept: 8. So $y = 4x + 8$.

15. A graph that goes downward means the output (gas) is getting smaller over time.

16. A reflection mirrors (flips) a figure across a line of reflection.

17. Two figures are congruent if one can be mapped onto the other using translations, reflections, and/or rotations — all rigid transformations.

18. Divide the image coordinates by the scale factor. $(9 \div 3, -15 \div 3) = (3, -5)$.

19. Any circle can be mapped onto any other circle by a translation (to align centers) followed by a dilation (to match radii).

20. $(2x + 5) + 3x + (x + 25) = 180$. Combine: $6x + 30 = 180$, $6x = 150$, $x = 25$.

21. $a^2 = 17^2 - 8^2 = 289 - 64 = 225$, so $a = 15$. (Note: $\sqrt{225} = 15$, so B and D are the same value, but B is simplified.)

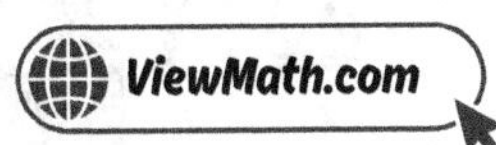

22 $d = \sqrt{(4-1)^2 + (5-1)^2} = \sqrt{9+16} = \sqrt{25} = 5.$

23 $V = \pi(6^2)(5) = \pi(36)(5) = 180\pi \ cm^3.$

24 $V = \frac{1}{3}(12 \times 5)(8) = \frac{1}{3}(60)(8) = \frac{480}{3} = 160 \ ft^3.$

25 Two variables may be associated due to a third variable or coincidence. Causation requires controlled experiments.

26 A straight line of best fit only works for data with a linear pattern. Curved data needs a different model.

27 Slope $= 0.4$ means 0.4 cm of growth per day. y-intercept $= 2$ means the initial height is 2 cm.

28 A two-way frequency table displays data for two categorical variables simultaneously, showing how frequently each combination occurs.

29 $P(A \ and \ B) = 0.6 \times 0.4 = 0.24$

30 $2^5 = 32.$ Each flip has 2 outcomes.

✅ Practice Test 8 — Answer Key

1 C   **2** A: ≈ 4.2, B: ≈ 6.3, C: ≈ 8.5 **3** ≈ 8.7 feet   **4** B   **5** -5 **6** C   **7** A

8 C   **9** B   **10** 16 and 20   **11** C   **12** C   **13** $y = x^2$

14 $y = 12x + 50$; $m = 12$ is the hourly wage, $b = 50$ is the base pay.   **15** Increasing, decreasing, increasing.

16 $(-a, -b)$ **17** $55°$ **18** C   **19** $\frac{3}{2}$ **20** C   **21** 13 ft   **22** B   **23** B   **24** 6 in

Find more at
ViewMath.com/IN-Grade8

 25 hours  B  $0.50 \times 0.20 = 0.10 = 10\%$ 30 B

5 Yes. Example: population growth over time may curve upward (exponential), showing a positive but nonlinear association

6 C 27 25 hours 28 B 29 $0.50 \times 0.20 = 0.10 = 10\%$ 30 B

💡 Time to Learn! 💡

Review the explanations below, **especially for the questions you missed.**

Understanding why each answer is correct builds stronger problem-solving skills.

Tip: Circle any questions you got wrong, then read their explanation carefully.

📖 Practice Test 8 — Detailed Explanations

1 Every integer n can be written as $\frac{n}{1}$, making it rational. Not every square root is irrational (e.g., $\sqrt{4} = 2$).

2 Side $= \sqrt{Area}$. $\sqrt{18} \approx 4.24 \approx 4.2$. $\sqrt{40} \approx 6.32 \approx 6.3$. $\sqrt{72} \approx 8.49 \approx 8.5$.

3 Side $= \sqrt{75}$. $8.6^2 = 73.96$ and $8.7^2 = 75.69$. Since 75 is between these, $\sqrt{75} \approx 8.7$.

4 By the negative exponent rule, $\frac{1}{a^n} = a^{-n}$, so $\frac{1}{2^4} = 2^{-4}$.

5 $(-5)^3 = -125$, so $\sqrt[3]{-125} = -5$.

6 Planet A is 1.5×10^8 km away. Ten times that is 1.5×10^9 km, which matches Planet D.

7 Hose A: 15 gal/min. Hose B: $\frac{100}{8} = 12.5$ gal/min. Hose A is faster.

8 A slope of 0 means there is no rise — the line is horizontal.

  Find more at
ViewMath.com/IN-Grade8

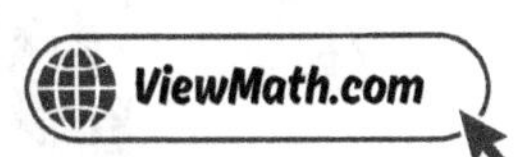

9 $\frac{6}{3} = 2$ and $\frac{10}{5} = 2$. Similar triangles on the same line always produce the same slope.

10 $x = y + 4$ and $x + y = 36$. Substitute: $(y + 4) + y = 36$, $2y = 32$, $y = 16$. Then $x = 20$.

11   Find the column where $f(x) = 15$. That column has $x = 4$.

12   Function R: when $x = 0$, $y = 1$. Function S: $b = 1$. Both start at 1, so the initial values are equal.

13   Any equation where x has an exponent other than 1 (such as x^2, x^3, $\frac{1}{x}$, $\sqrt{x}$) is nonlinear.

14   Slope: $62 - 50 = 12$ (hourly rate). y-intercept: 50 (base pay before any hours). Equation: $y = 12x + 50$.

15   Filling: the water level rises. Drinking: the level drops. Refilling: the level rises again.

16   Reflect over x-axis: $(a, b) \rightarrow (a, -b)$. Then reflect over y-axis: $(a, -b) \rightarrow (-a, -b)$. This is the same as a $180°$ rotation.

17 $\angle R = 180 - 72 - 53 = 55°$. Since $\angle R$ corresponds to $\angle U$, $\angle U = 55°$.

18   The rule $(x, y) \rightarrow (-y, x)$ is a $90°$ counterclockwise rotation around the origin.

19 $\frac{9}{6} = \frac{3}{2}$, $\frac{12}{8} = \frac{3}{2}$, $\frac{15}{10} = \frac{3}{2}$. The scale factor is $\frac{3}{2}$.

20   By the Exterior Angle Theorem, y equals the sum of the two non-adjacent interior angles: $y = 55 + 75 = 130°$.

21 $d = \sqrt{12^2 + 5^2} = \sqrt{144 + 25} = \sqrt{169} = 13$ ft.

Find more at
ViewMath.com/IN-Grade8

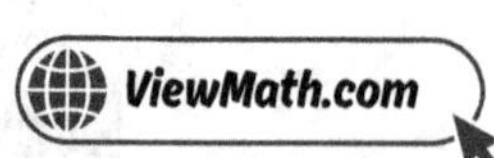

22. The distance formula uses the horizontal and vertical distances as legs of a right triangle and applies $a^2 + b^2 = c^2$.

23. $V = \frac{4}{3}\pi r^3 = \frac{4}{3}(3.14)(27) = \frac{4}{3}(84.78) = 113.04 \ cm^3$.

24. $48 = \frac{1}{3}(s^2)(4)$, $48 = \frac{4s^2}{3}$, $144 = 4s^2$, $s^2 = 36$, $s = 6 \ in$.

25. An association can be positive (both increase) and nonlinear (the trend is curved, not straight).

26. The line of best fit does not need to pass through any actual data point. It just needs to be close to most of them.

27. Set $y = 0$: $0 = -2x + 50$, $2x = 50$, $x = 25$.

28. $\frac{22}{40} = 0.55 = 55\%$.

29. Spins are independent. $P = 0.50 \times 0.20 = 0.10$.

30. $P(8, 3) = 8 \times 7 \times 6 = 336$. Order matters.

✅ Practice Test 9 — Answer Key

1 C	2 B	3 B	4 A	5 B	6 B	7 C	8 8 meters per minute
9 B	10 7 T-shirts and 8 hats	11 C	12 A	13 D	14 B	15 C	16 B
17 A	18 B	19 B	20 A	21 C	22 B	23 B	24 B	25 C	26 C
27 B	28 A	29 B	30 $26 \times 25 \times 10 \times 9 = 58,500$

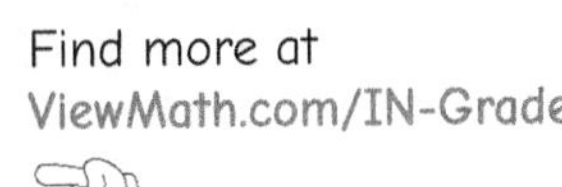

Find more at
ViewMath.com/IN-Grade8

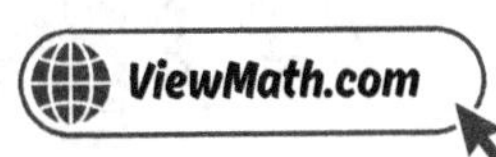

> 💡 **Time to Learn!** 💡
>
> Review the explanations below, **especially for the questions you missed**.
>
> Understanding why each answer is correct builds stronger problem-solving skills.
>
> **Tip:** Circle any questions you got wrong, then read their explanation carefully.

📖 Practice Test 9 — Detailed Explanations

1 $1.41421356\ldots$ is the decimal expansion of $\sqrt{2}$, which is non-repeating and non-terminating, indicating an irrational number.

2 $3^2 = 9$ and $4^2 = 16$. Since $9 < 10 < 16$, we have $3 < \sqrt{10} < 4$.

3 $\sqrt{2} \approx 1.414$, so $3\sqrt{2} \approx 4.243$. This is between 4 and 5.

4 $3^{-2} = \frac{1}{3^2} = \frac{1}{9}$.

5 $7^2 = 49$ and $8^2 = 64$. Since $49 < 50 < 64$, we have $7 < \sqrt{50} < 8$.

6 $(2.4 \times 10^3)(5 \times 10^2) = 12 \times 10^5 = 1.2 \times 10^6$.

7 $k = \frac{12}{8} = 1.5$. At $x = 20$: $y = 1.5 \times 20 = 30$.

8 $m = \frac{40-0}{5-0} = \frac{40}{5} = 8$ meters per minute.

9 A line through the origin has $b = 0$. Only $y = -3x$ has no constant term added.

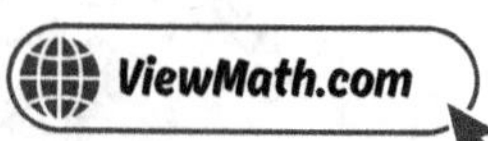

10 $t + h = 15$ and $12t + 8h = 148$. From first: $h = 15 - t$. Substitute: $12t + 8(15 - t) = 148$, $4t + 120 = 148$, $4t = 28$, $t = 7$. Then $h = 8$.

11 $f(-3) = -(-3) + 8 = 3 + 8 = 11$.

12 The rate of change is the slope. Function A has slope 5 and Function B has slope 2. Since $5 > 2$, Function A has the greater rate of change.

13 $y = \sqrt{x}$ involves a square root of the variable, so it is nonlinear. The others all fit $y = mx + b$.

14 Slope: $\frac{11-5}{2-0} = \frac{6}{2} = 3$. The y-intercept from $(0, 5)$ is $b = 5$. So $y = 3x + 5$.

15 A function can decrease while staying positive (e.g., dropping from 10 to 5). "Decreasing" means the values get smaller, not that they are negative.

16 Translations preserve all side lengths. The longest side is still 8 cm.

17 Both translation and rotation are rigid transformations. The image is congruent to the original regardless of the shape.

18 The 90° CCW rule is $(x, y) \rightarrow (-y, x)$. So $(-3, 5) \rightarrow (-5, -3)$.

19 When $k = 1$, every point stays in the same place. The image is congruent (identical in size and shape) to the original.

20 $\angle C = 180 - 45 - 85 = 50°$. All three angles are less than 90°, so the triangle is acute.

21 $9^2 + 40^2 = 81 + 1600 = 1681 = 41^2$. The converse of the Pythagorean Theorem confirms it's a right triangle.

22 $d = \sqrt{3^2 + 4^2} = \sqrt{9 + 16} = \sqrt{25} = 5$.

Find more at
ViewMath.com/IN-Grade8

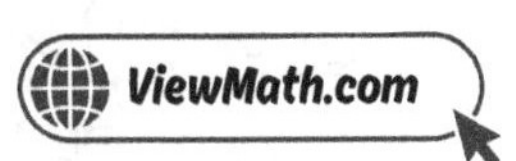

23. $80\pi = \frac{1}{3}\pi(16)h$. Divide both sides by π: $80 = \frac{16h}{3}$. Multiply by 3: $240 = 16h$. Divide: $h = 15$ cm.

24. $V = \frac{1}{3}Bh = \frac{1}{3}(40)(12) = 160\ m^3$.

25. The dots go from lower-left to upper-right in a roughly straight pattern. This is a positive linear association.

26. $y = 1.5(4) + 2 = 6 + 2 = 8$.

27. Residual $=$ actual $-$ predicted $= 8 - 6 = 2$.

28. Of 25 who did not study, 10 passed. $\frac{10}{25} = 0.40 = 40\%$.

29. $P = \frac{1}{2} \times \frac{1}{2} \times \frac{1}{2} = \frac{1}{8}$.

30. Letters: 26×25 (no repeat). Digits: 10×9 (no repeat). Total: $26 \times 25 \times 10 \times 9 = 58,500$.

✅ Practice Test 10 — Answer Key

 B    B    B    C    C   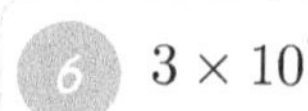 3×10^7  12 cups    3    C

 B    9    B    Differences are all 4; the function is linear.    B    B

 $(-5, -3)$  C    $(-5, -2)$  A    B    B    B    $\approx 226.08\ in^3$

 C    A

26. Outliers can pull the line away from the overall trend, making it less representative of most data points.

27. B   28. B   29. C   30. B

Find more at
ViewMath.com/IN-Grade8

> ## 💡 Time to Learn! 💡
>
> Review the explanations below, **especially for the questions you missed**.
>
> Understanding why each answer is correct builds stronger problem-solving skills.
>
> **Tip:** Circle any questions you got wrong, then read their explanation carefully.

📖 Practice Test 10 — Detailed Explanations

1. 5 is not a perfect square, so $\sqrt{5}$ is irrational and belongs in the Irrational region.

2. $\sqrt{6} \approx 2.449$. Point B is located at approximately 2.45 on the number line, which is the closest to $\sqrt{6}$.

3. $\pi \approx 3.14$, so $\pi^2 \approx 3.14 \times 3.14 = 9.8596 \approx 9.9$.

4. A negative exponent flips the fraction: $\left(\frac{3}{4}\right)^{-2} = \left(\frac{4}{3}\right)^2 = \frac{16}{9}$.

5. $\sqrt{2}$ cannot be expressed as a fraction of two integers, so it is irrational.

6. $\frac{7.2 \times 10^{12}}{2.4 \times 10^5} = \frac{7.2}{2.4} \times 10^{12-5} = 3 \times 10^7$.

7. $k = \frac{2}{5}$ cups per cookie. For 30 cookies: $\frac{2}{5} \times 30 = 12$ cups.

8. $m = \frac{11-(-1)}{7-3} = \frac{12}{4} = 3$.

9. Through the origin, $b = 0$. Slope $= \frac{12}{3} = 4$. Equation: $y = 4x$.

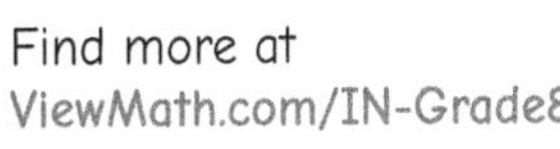
Find more at
ViewMath.com/IN-Grade8

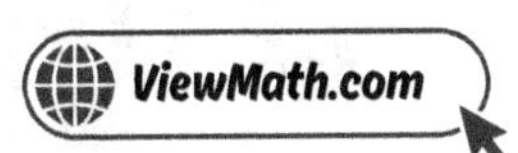
ViewMath.com

10. $p + n = 8$ and $2p + 5n = 25$. From first: $p = 8 - n$. Substitute: $2(8 - n) + 5n = 25$, so $16 + 3n = 25$, $3n = 9$, $n = 3$.

11. From the graph, $f(1) = 3$ and $f(4) = 6$. So $f(1) + f(4) = 3 + 6 = 9$.

12. Set $2x + 100 = 8x + 10$. Subtract $2x$: $100 = 6x + 10$. Subtract 10: $90 = 6x$. Divide: $x = 15$.

13. $1 - (-3) = 4$, $5 - 1 = 4$, $9 - 5 = 4$, $13 - 9 = 4$. The constant difference of 4 means the rate of change is constant, so the function is linear.

14. Starting amount $b = 120$. Spending \$15/day means losing money, so $m = -15$. The function is $y = -15x + 120$.

15. A steeper distance-time graph means more distance covered per unit of time — that means greater speed.

16. Reflecting over the y-axis flips the sign of the x-coordinate: $(5, -3) \rightarrow (-5, -3)$.

17. Two circles with the same radius are always congruent — one can be translated to overlap the other.

18. 90° CCW: $(x, y) \rightarrow (-y, x)$. So $(-2, 5) \rightarrow (-5, -2)$.

19. Perimeter scales by the same factor. $36 \times \frac{1}{3} = 12$ cm.

20. Co-interior angles are supplementary: $180 - 135 = 45°$.

21. $c^2 = 6^2 + 8^2 = 36 + 64 = 100$, so $c = 10$.

22. The x-coordinates are the same, so the distance is the vertical difference: $|7 - (-5)| = 12$.

Find more at
ViewMath.com/IN-Grade8

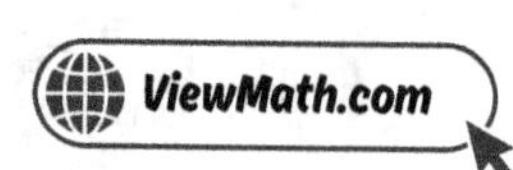

23 $V = 3.14 \times 9 \times 8 = 226.08 \ in^3.$

24   Prism volume $= 6 \times 6 \times 10 = 360 \ in^3.$ Pyramid volume $= \frac{360}{3} = 120 \ in^3.$

25   Clusters are groups of points that bunch together, while outliers are isolated far from the trend. Both can appear in the same scatter plot.

26   An outlier is an unusual point. Including it would distort the line and give a worse fit for the majority of the data.

27   The y-intercept ($b = 10$) is the predicted value when $x = 0$, meaning the predicted score with no study time.

28   Boys: 6 summer vs. 3 winter. Girls: 4 summer vs. 5 winter. Boys lean toward summer, girls lean toward winter.

29 $P(6) = \frac{1}{6}.$ Independent: $P = \frac{1}{6} \times \frac{1}{6} = \frac{1}{36}.$

30 $\binom{5}{2} = \frac{5!}{2! \cdot 3!} = \frac{120}{2 \cdot 6} = 10.$

Well done checking your answers!

Keep practicing to strengthen your skills.

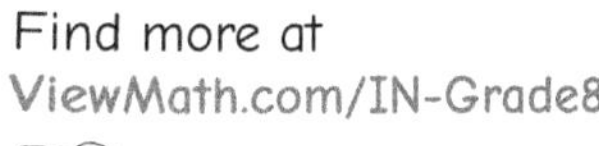